HAZARDOUS MATERIALS AND WASTE MANAGEMENT

HAZARDOUS MATERIALS AND WASTE MANAGEMENT

A Guide for the Professional Hazards Manager

by

Nicholas P. Cheremisinoff
Paul N. Cheremisinoff

National Association of Safety & Health Professionals
and
Fairleigh Dickinson University

np NOYES PUBLICATIONS
Park Ridge, New Jersey, U.S.A.

Library of Congress Catalog Card Number 94-38698
ISBN: 0-8155-1372-0
Printed in the United States

Published in the United States of America by
Noyes Publications
Mill Road, Park Ridge, New Jersey 07656

Transfered to Digital Printing, 2011

Printed and bound in the United Kingdom

Library of Congress Cataloging-in-Publication Data

Cheremisinoff, Nicholas P.
     Hazardous materials and waste management : a guide for the
professional hazards manager / by Nicholas P. Cheremisinoff, Paul N.
Cheremisinoff.
          p.   cm.
     Includes index.
     ISBN 0-8155-1372-0
     1. Hazardous wastes--Management. 2. Hazardous substances--Safety
measures. I. Cheremisinoff, Paul N. II. Title.
TD1030.C49   1995
363.72'875--dc20
                                                     94-38698
                                                       CIP

ABOUT THE AUTHORS

Nicholas P. Cheremisinoff is the Director of the Center of Environmental Management at Fairleigh Dickinson University. He is also advisor and executive director of the National Association of Safety & Health Professionals, 457 Highway 79, Morganville, NJ 07751, which provides a national certification and registration program for hazard materials handling specialists and environmental managers. Dr. Cheremisinoff has had 20 years of industry and applied research experience, and is an internationally recognized expert, having authored, co-authored and edited over 100 engineering textbooks. He received his B.S., M.S., and Ph.D. degrees in chemical engineering from Clarkson College of Technology.

Paul N. Cheremisinoff is a registered Professional Engineer and is full professor in the Civil and Environmental Engineering Department of New Jersey Institute of Technology. He is internationally recognized as an expert in pollution control and hazardous materials handling and remediation technolgies, with over 40 years of industrial and applied research experience. He is a consultant to numerous fortune 500 corporations and government agencies, and has authored, co-authored, or edited over 300 engineering textbooks. He is also editor of The National Environmental Journal, which reaches over 90,000 environmental managers and specialists.

PREFACE

The management of hazardous materials and industrial wastes is complex, requiring a high degree of knowledge over very broad technical and legal subject areas. Hazardous wastes and materials are diverse, with compositions and properties that not only vary significantly between industries, but within industries, and indeed within the complexity of single facilities. Proper management not only requires an understanding of the numerous and complex regulations governing hazardous materials and waste streams, but an understanding and knowledge of the treatment, post–treatment and waste minimization technologies. In fact, today's environmental manager must face working within 12 environmental management arenas, all of which may be applicable regardless of the size of the operation or business.

This volume has been written as a desk reference for the Professional Hazards Manager (PHM). The PHM is a qualified environmental manager that has the responsibility of ensuring that his or her facility or division within the corporation is in compliance with environmental statues and regulations, as well as participating in the selection of technologies and approaches to remediation, pollution control and in implementing waste minimization practices. These decisions require knowledge and understanding of the federal, state and local environmental regulations, a working knowledge of the best available technologies and their associated cost. This volume provides an overview of both the technology and compliance requirements that will assist environmental managers in addressing facility management of hazardous wastes, pollution control, and waste minimization. The book has been designed in part as a study guide to help prepare qualified individuals for the national certification and registration program of Professional Hazards Managers conducted by the National Association of Safety & Health Professionals and other organizations including the Hazard Materials Control Resources Institute (HMCRI) and Fairleigh Dickinson University.

Nicholas P. Cheremisinoff, Ph.D.
Paul N. Cheremisinoff, P.E.

TABLE OF CONTENTS

CHAPTER 5. HAZARDOUS MATERIAL AND SLUDGE TREATEMNT
AND DISPOSAL

CHAPTER 6. RECOVERY SYSTEMS FROM WASTES DISPOSAL

CHAPTER 7. WASTE MINIMIZATION

CHAPTER 8.   WORKING WITH HAZARDOUS MATERIALS

CHAPTER 9.   ESTIMATING RELEASES TO THE ENVIRONMENT

CHAPTER 10. REGULATORY COMPLIANCE: AN OVERVIEW OF WORKER
PROTECTION AND RIGHT-TO-KNOW

CHAPTER 11. REGULATION OF HAZARDOUS WASTES

CHAPTER 12. MANAGING ENVIRONMENTAL COMPLIANCE

CHAPTER 1

HAZARDOUS WASTE MANAGEMENT
AND POLLUTION CONTROL

INTRODUCTION AND OVERVIEW

Facilities that generate, transport, transfer or dispose of hazardous wastes require substantial construction, operating costs and have become extremely expensive and complex. Handling and disposal of hazardous waste and attending requirements have been and promise to be substantially expanded and subject not only to technology needs but to the hazardous waste regulatory environment.

Improper **hazardous waste management** is a most serious environmental problem facing us today. If anyone needs convincing that this is the case, we need only to consult the newspapers. Hardly a week goes by without a major new revelation of a hazardous waste mishap. Attention can be focused to events related to waste site conditions, waste disposal, air pollution concerns, water contamination, garbage disposal, to health effects of people having significant chromosomal aberrations, genetic damage, warnings of present and future health problems including cancer, birth defects, spontaneous abortions and other problems requiring further follow up research dictating more monitoring requirements. And problems are not just limited to the thousands of damage incidents already

documented. There are numerous dangerous hazardous waste sites, and until cleaned up or remedied are catastrophes waiting to happen.

RCRA (Resource Conservation and Recovery Act) enacted in 1976 established a comprehensive, national regulatory policy for managing hazardous wastes from **cradle to grave**. EPA under its statutory authority authorizes states to administer and enforce hazardous waste programs no less stringent than the RCRA regulations promulgated by EPA. EPA has promulgated methods for determining whether solid wastes must be regulated as hazardous. Hazardous wastes include any solid waste identified and deemed hazardous and set out as hazardous regardless of their sources; and wastes that are hazardous from particular sources and processes; and wastes that are hazardous regardless of their sources; and wastes that are hazardous if from particular sources; as well as wastes of discarded chemical products.

In identifying hazardous wastes, wastes are analyzed for certain characteristics at specified levels of hazards. Characteristics considered as classifiers of hazardous waste include: **toxicity; ignitability; reactivity; corrosivity.**

1

Environmental regulations enacted in 1984 and amendments to the Superfund program discourage landfilling of wastes in favor of remedial methods that will treat or destroy wastes. New alternative technologies seek to destroy, stabilize, or treat hazardous wastes by changing their chemical, biological or physical characteristics. Increasing environmental concerns and an increasingly stringent regulatory climate should bring about changes in the hazardous waste cleanup business. The wide range of problems encountered requires specific solutions often at high costs.

Control, remediation, and cleanup must be balanced based on the effectiveness of technological techniques and the ultimate action considered in light of the specific situation. Tighter regulations, more environmental concerns, and changes in control and treatment technologies make the hazardous waste cleanup business more and more complicated. New and emerging technologies offer a better approach to solving these problems. Landfilling hazardous wastes and the potential for landfill leakage and hazardous materials released into the environment is and promises to be a problem well into the future.

Notwithstanding the existence of regulations, industry has the opportunity to control the pattern of hazardous waste management. In addition to protecting the public health and the environment, industry also protects itself from laws and governmental actions which they may find unreasonable.

The hazardous waste problem can be divided into two parts. The first problem can be considered a technical problem dealing with the tens of thousands of generators of hazardous wastes producing millions of tons of such wastes with ever increasing amounts being generated each year. The second problem comes to light in the thousands of pages of regulations we have to live with. Both the technical and regulatory aspects present the very serious and difficult problems faced. **Table I** provides a partial listing of typical industrial wastes.

TABLE I.

TYPICAL INDUSTRIES AND THEIR HAZARDOUS WASTES INCLUDE:

INDUSTRY	TYPICAL WASTE	TYPICAL FORM
Agriculture	Organics, pesticides, herbicide residues	Sludge
Chemical	Acids, alkalies, metals, organics	Liquid, sludge solids
Electronics/ electroplating	Heavy metals, cyanides	Sludge
Machinery	Oils	Liquid
Metals refining	Heavy metals, cyanides	Liquid sludge
Municipal waste treatment	Organics, heavy metals	Sludge
Petroleum	Heavy metals, acids, alkalis	Liquid, solid, sludge
Paint	Heavy metals	Liquid, sludge
Wood Processing	Heavy metals, organics	Sludge, solid

The Complex Web of Laws and Regulations

The two principal federal laws that try to control hazardous wastes are the Resource Conservation & Recovery Act (RCRA) and the Comprehensive Environmental Response, Compensation and Liability Act (CERCLA/Superfund). The U.S. Environmental Protection Agency (EPA) administers both and is designed to ensure that hazardous wastes are not disposed of in such a way as to harm human health or the environment.

RCRA, which passed in 1976, was strengthened by the Hazardous and Solid Waste Amendments of 1984. It provides "cradle to grave" management for more than 450 chemical wastes listed in the law as hazardous. Specific permits for treatment, storage, disposal of hazardous wastes, and manifest tracking ensures proper handling of wastes generated.

Superfund, provides for the cleanup of old waste sites, landfills, waste lagoons/ponds, warehouses where hazardous chemicals have been abandoned. EPA has currently listed more than 1000 abandoned sites on its National Priority List.

Land-ban rules by Congress listed land disposal deadlines for specific wastes considered hazardous and ordered EPA to ban land disposal of any of these after their deadlines unless wastes have first been made non hazardous. Materials on the list include dioxins, certain organic solvents, various chemicals called the **California List** and a long list of other industry waste streams. One objective has been to drastically reduce the volume of these chemicals or completely stop their generation. The land ban of specific, listed wastes and waste streams was to be in effect by May 1990.

Has the time finally come when our disposable, use-it-once and throw-it-away society must pay the piper at last in forms of vastly greater regulation and control over how we handle the wide range of toxic and hazardous waste materials. These are materials that we really should not have been so cavalier about in the first place. Industry's potential exposure to liability for hazardous waste disposal is horrendous today and regulations don't alleviate that liability one bit. It is even conceivable that waste products we think benign today could prove hazardous tomorrow. In the current legal climate, state-of-the-art technology or good faith practice is a risky, if not inadequate defense.

Treatment Technologies for Hazardous Wastes

Manufacturing operations in most cases still generate waste products that need to undergo treatment to destroy or render them environmentally acceptable. The numerous technologies applicable to hazardous wastes can be subdivided into chemical, physical, or biological classes of which many are widely used to manage hazardous wastes. Use of combinations of such treatment technologies are widely used for cost-effective management.

Chemical Treatment

Chemical treatment transforms hazardous waste streams by a number of successful technologies which include:

- Neutralization
- Color Removal
- Precipitation

- Disinfection
- Coagulation and Flocculation
- Ion Exchange
- Oxidation and Reduction
- Stabilization

Physical Treatment

Physical treatment consists of a wide variety of separation methods that have been common to industry for many years and includes:

- Screening
- Filtration
- Sedimentation
- Sorption
- Clarification
- Evaporation/Distillation
- Centrifugation
- Stripping
- Flotation
- Membrane Technologies

Biological Treatment

Biological treatment has proven to be a cost-effective and efficient way for removal of hazardous wastes from contaminated wastewater; groundwater; landfill leachate and contaminated soil. Technologies include:

- Aerobic systems
- Anaerobic systems

Biodegradability is pollutant and system specific and particular compounds may not degrade in some systems. Microorganisms for biological treatment can be categorized as to oxygen utilization and may be classified as heterotrophic or autotrophic, depending on their nutrient source.

Aerobic systems and **micro-organisms** are most commonly used for industrial wastewater and usually for treatment of strong organic wastes or sludges from aerobic processes. **Anaerobic microbes** use oxygen that is combined chemically with other elements in compounds such as nitrates, carbonates, and sulfates.

Incineration/Thermal Processes

Incineration offers a disposal technology for substances having a high heat release potential, particularly hydrocarbons. Reduction in volume and weight for bulk wastes make management simpler and provides advantages over other waste treatment processes, particularly landfilling. Incineration of hazardous organic wastes affords efficient destruction with usually controllable air emissions.

Even though hazardous waste incineration has many potential advantages, it also has its drawbacks which include:

- high initial capital investment,
- air emission control requirements,
- high maintenance requirements.

Hazardous wastes incineration has developed from open-pit burning to highly efficient and sophisticated systems. New facilities and complex designs and operations have resulted in higher costs. In addition to technical and economic factors, siting of waste incinerators is a highly controversial issue. Emerging technologies in the incineration/ thermal destruction area include the use of oxygen; slagging and vitrification processes and pyrolysis.

The Europeans are handling individual waste problems in a variety of ways. Some common threads indicate **incineration** is the preferred method for waste destruction, with bottom ash or slag being landfilled. Incinerator capacity as in the U.S. is very tight and leads to environmental concerns in considering facility expansion. European environmental concerns were originally sparked by the Seveso, Italy accident in 1976 which sprayed 2,3,7,8-tetra-chlorodibenzo-p-dioxin (TCDD) into the environment.

Incineration is increasingly being turned to for destruction of hazardous wastes and sludges. Industry responsible for much of the hazardous wastes generated is no longer able to rely on landfills or such options as injection wells for wastes disposal. Regulations once fully implemented will require new and changing disposal and treatment strategies. It is estimated that hazardous wastes incineration will double in the 1990s and may even increase five fold by the year 2000. Estimates are that 12 million metric tons will be treated by incineration technology in the year 2000, representing only a small fraction (5-6%) of the hazardous wastes total produced in the U.S. Hazardous wastes incineration will represent a $6-billion equipment market annually by the turn of the century.

Disposal practices have heretofore relied largely on EPA-certified landfills. This latter option economically favored ($25-$200 per ton of waste) is far below the second most widely favored alternative, incineration ($500-$2000 per ton). Environmental concerns; public and political pressure has restricted new landfills as well as closing old ones due to unsuitable geology and violations. These factors have made landfilling costs rise substantially. The re-authorized Resource Conservation and Recovery Act (RCRA) prohibits burying of certain untreated wastes, including solvents, halogenated

organics and dioxin. The ideal incineration system minimizes fuel consumption and/or maximizes energy recovery while converting wastes to environmentally acceptable forms. EPA requires incinerators to achieve at least a 99.99% <u>Destruction Efficiency</u> of the Principal Organic Hazardous Constituents (POHCs) present in the waste.

Hazardous Wastes Site Cleanup with On-Site Technologies

CERCLA (The Comprehensive Environmental Response, Compensation, and Liability Act/the Superfund Law) began a broad national program in the U.S. to clean up chemically contaminated sites which are estimated to number in the tens of thousands. Problems attendent with such sites include: residual disposal; health and safety during cleanup; financial liabilities. Costs are the prime consideration, with multimillion dollar cleanups common. Superfund sites are only a part of the problem and costs promise to become increasingly greater unless better methods become available in the future.

Chemical waste cleanup problems present an enormous variety of problems. Costly site specific problems and their many combinations to treat the wide range of waste materials includes the following technologies for on-site treatment:

Bioremediation - has been demonstrated as effective for halogenated aliphatics; nitrated compounds; heterocyclics; polynuclear aromatics; polar nonhalogen organics. This technology may be potentially effective for nonpolar halogenated aromatics, PCBs, and dioxins. Biotreatment is not effective on biotoxic wastes such as heavy metals.

Immobilization - has been demonstrated as effective for heavy metals.

Solvent extraction - has been shown effective for simple non-halogenated aromatics. Other methods indicated may be potentially effective in certain situations. Solvent extraction methods may also be classed as soil-washing.

Thermal destruction - has been indicated as effective for a wide range of wastes except for metals. Thermal processes now under consideration though not widely practiced or proven include use of oxygen and slagging/vitrification and pyrolysis.

Thermal desorption (low temperature) - has shown effectiveness for simple non-halogenated and halogenated organics.

Other on site technologies include neutralization/volatilization via soil aeration/vacuum extraction/containment.

Technical solutions for cleanup sites may also involve either long term containment or immediate attempts for treatment to destroy the waste or make it environmentally acceptable. **Containment technologies** are usually only interim solutions. Containment systems for hazardous wastes sites include:

- **Groundwater Barriers**
  — Slurry walls
  — Grout curtain
  — Vibrated beam
  — Sheet pile
  — Block displacement

- **Groundwater pumping**

- **Subsurface drains**
- **Runoff controls**
- **Surface and seal caps**

- **Solidification and stabilization**
- **Encapsulation.**

Fixation and Solidifications

As more attention is focused on incineration and thermal treatment, the question increasingly arises as to the safe disposition of residues.

Stabilization of wastes is essentially a pretreatment process before solidification. These are processes that physically stabilize liquid/semi-liquid waste residues. This term is often confused with **fixation** and should be clearly defined and understood when used. **Chemical fixation** on the other hand can be defined as the chemical technology to immobilize, insolubilize, detoxify a waste, rendering the waste or its components less hazardous. Heavy metals ion-exchange; sorption processes on flyash or carbon materials are examples. Solidification by itself may not reduce hazard potential, but will reduce solubility and the effective exposed surface to the environment. Dewatering; filtration; drying are further examples of the solidification process and do not involve chemical reaction.

Nonchemical methods of solidification include dewatering; adsorption and vegetative or crop stabilization. Water may be removed by filtration or drying and is applicable in sludge treatment and handling. Admixing or adsorption of waste residues has been practiced and studied. Sorbents vary from high capacity adsorbents to by-product wastes themselves such as flyash and municipal refuse. Sewage and mine wastes have been disposed of by spraying, plowing, spreading into soil. Vegetative stabilization involves the growing of plants whose roots stabilize the soil. In this latter case the waste is compatible with vegetation grown and therefore may contain plant nutrients.

Final land use resulting from such treatment is a factor to consider.

Encapsulation processes employing organic systems has great potential. Some systems that have been used or considered include:

— Asphalt (bitumen)
— Polyolefin encapsulation
— Epoxy resins
— Polyester resins
— Polybutadiene
— Urea-formaldehyde
— Acrylamide gels.

Such systems are based on polymerization and may be inhibited by the presence of water or other constituents. Inorganic systems have involved portland cement; portland cement-flyash; lime-portland cement; portland cement-sodium silicate. Inorganic systems as a rule are lower in cost than organic solidification. Obviously the waste nature affects the process choice.

Combining wastes with solidification agents can be done considering the following options:

- In-drum
  — On-site
  — Off-site
- In-situ
  — Deep soil mixing
  — Detoxification
  — Backhoe mixing
  — Grouting
- Plant mixing
  — Pug mill
  — Mobile plants
  — Special processes

- Area mixing

GROUNDWATER CONTAMINATION

Groundwater is the source of approximately 20 percent of the U.S. domestic, agricultural and industrial water supply and in many parts of the country is the only dependable water supply. Groundwater is stored in aquifers (geological formations of permeable saturated rock, sand or gravel zones) and are recharged as atmospheric precipitation or surface waters drains into them. Wherever or whenever the presence of chemical contamination in groundwater exists, the result is a problem for human health and the environment. The presence of many chemicals present serious and substantial health risks, even in low concentrations and may have mutagenic, teratogenic, carcinogenic properties. Sources of contamination are many and varied and as chemical usage and industrial activity increase so has the potential for contamination.

Contaminants of concern include organic solvents, petroleum products, gasoline, pesticides and nitrates. Waste disposal practices present possibly the greatest threat of groundwater contamination. Landfill disposal of hazardous wastes was long accepted as a suitable disposal practice. There are nearly 20,000 abandoned and uncontrolled hazardous waste sites in the U.S. alone with many of these indicating some degree of groundwater contamination. Additionally, there are nearly 100,000 landfills in the U.S. disposing of nonhazardous/household wastes and represent a potential source of groundwater contamination from their leachate production.

STORAGE TANKS

Environmental hazards involving oil transport, transfer and storage, have caused numerous instances of drinking water and water contamination. Storage tanks (whether under or above ground) as well as the transportation of petroleum products/chemicals are regulated by environmental law. Control of storage tanks will depend on tank contents, size, design and location. Failure to comply with some of the complex regulations such as registration, reporting, design, operating standards and monitoring can subject an owner to significant penalties.

The scope of the problem can be illustrated by way of New York State as an example. Existing facilities required to register with the New York State Department of Environmental Conservation (DEC) in 1986 numbered some 132,000 tanks containing nearly 4,400 million gallons. New facilities require registration before being placed into service. Another example is the State of New Jersey, which has over 80,000 registered commerical and industrial underground storage tanks (USTs). Many of these tanks either leaked due to corrosion, or their piping distribution system's leaked, or many years of overfilling resulted in contaminating the soil, ultimately causing groundwater contamination due to leachate formation. To correct this, UST regulations require upgrading to double wall tankage and piping, as well as the use of overfill ports, automatic leak detection and alarms, and vapor extraction lines. An industry which many claim has suffered from the UST regulations is the gasoline service stations, principally because serious contamination is often found in the soil and water.

Some of the requirements for operating such facilities today include:

Operators of above ground storage tanks must conduct monthly visual inspections and clean tanks out periodically (i.e. every ten years) removing bottom sludges; structural integrity and tightness testing.

Tanks temporarily out of service (>30 days) must be drained to the lowest drawoff point; full lines, gauge openings must be capped or plugged. Tanks permanently out of service must be emptied of liquid/sludge/vapors and either be removed or filled with solid inert material such as sand or concrete slurry.

New underground storage tanks must either be made of fiberglass reinforced plastic, cathodically protected steel for corrosion protection between soil and steel contact, or steel clad with fiberglass reinforced plastic. Secondary containment provided as either double walled tanks; a vault; a cut-off wall; impervious underlayment.

For monitoring, underground tanks must be equipped with double walls with monitoring provisions of interstitial space; in-tank monitoring systems; or observation wells.

New piping systems should be cathodically protected iron or steel; fiberglass reinforced plastics or equivalent non-corrodible materials.

Storage facilities handling hazardous substances are also subject to release reporting requirements, where there are releases of a reportable quantity (i.e., quantities in excess of numbers specifically assigned in regulations), or unknown quantities of hazardous substances. As an example, the reportable quantity (RQ) for gasoline is one gallon.

RCRA amendments in 1984 and 1986 established a federal program for underground storage tanks (USTs). These amendments apply to tanks whose volume is 10% or more below ground, containing petroleum, petroleum products, or hazardous substances. Exemptions include hazardous waste/wastewater treatment tanks that are part of treatment facilities regulated under the Clean Water Act and equipment containing regulated substances for facilities such as hydraulic and electrical equipment tanks. USTs are subject to an exhaustive and comprehensive set of requirements. Regulations governing underground and above-ground storage tanks are numerous, complex and developing. One area where they are currently evolving is in the regulation of home-owner fuel oil tanks.

OILY WASTES AND OIL SPILLS

A problem of concern to many plant and industrial engineers is the proper disposal of unwanted waste oils and wastewaters containing petroleum products. A wide range of lubrication wastes include oils, emulsifiers, acids, metals, dissolved and suspended solids, organic and inorganic salts. Oil emulsions can be particularly troublesome to control and even when treated effluents can contain substantial quantities of soluble oil.

There are several methods that can be used to reduce pollution and overall abatement costs before end-of-pipe treatment. The best opportunity to reduce wastes is during engineering design. Once a facility is in operation, pollution control may become necessary, difficult and expensive. Any first step must include review of the operation to reduce the waste. The following physical and chemical methods are employed in waste oil recovery:

Gravity differential separation - separation alone may remove water and settleable solids, with free oil being recovered.

Vacuum filtration - precoat vacuum filtration incorporates a layer of filter aid on a vacuum rotary drum and is an established method for treatment of petroleum refining waste.

Acid treatment - sulfuric acid treatment can break emulsions and separate saturated napthenic/paraffinic molecules.

Temperature - heat treatment can be used alone or in conjunction with gravity separation and/or acid treatment to break emulsions for oil separation.

Centrifugation - centrifuging as a materials separation method for oils has been successful.

Chemical treatment - chemical methods for emulsion breaking are widely used and include reaction with salts of polyvalent metals, salting out of alkali soaps and destruction of emulsifying agents.

Flocculation and sedimentation - chemical flocculation can break emulsions. Materials such as clay and flyash have been used in conjunction with coagulants to effectively separate sludges.

Extraction - certain compounds can be separated by selective extraction.

Agitation - some emulsions may be broken by vigorous agitation.

Controlling Oil Spills

The oil spill problem has received much attention in recent years as a result of environmental concerns and has become a relatively high attention getting factor when it does occur. Oil is in constant transit with problems arising from spill problems at sea to simple transfer operations. The problem of oil pollution itself is complex and must take into account the interaction of oil, water, wind, etc. Attempts are made immediately on detecting oil spills to contain the oil and then collect for further treatment and or disposal. Collection or containment devices include booms and skimmers and in some case adsorbent materials such as straw.

Hazardous Wastes Transportation

It is estimated that over 500,000 shipments of hazardous materials move through the United States per day. Combine this with the multitude of stationary sources which manufacture, store and consume hazardous materials and we see that a significant problem exists.

Historically transportation of hazardous materials has been regulated by federal law since 1866 which covered shipments of explosives and flammable materials. After the Civil War, rail shipments of explosives were regulated by uncodified statutes and contracts based on English common law.

The Interstate Commerce Commission (ICC) was established in 1887 and until the formation of the Department of Transportation (DOT) in 1966 was the primary transportation regulatory authority. In 1975, the Hazardous Materials Transportation Act (HMTA) was passed to improve regulatory and enforcement activities and set regulations applicable to hazardous materials transportation.

It has been estimated that 270 million metric tons of hazardous materials are generated in the U.S. yearly. While most of this waste is disposed of on-site, wastes shipped off-site are usually transported by truck. Transportation of these materials is regulated by both HMTA and RCRA.

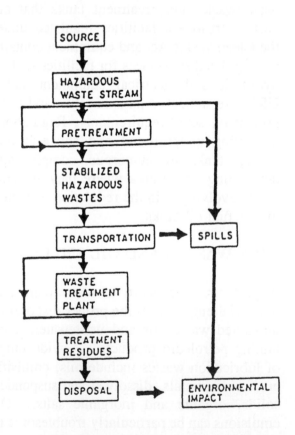

Basic steps in handling, treating and disposing of hazardous wastes.

Regulations cover the following:

- Identification, listing, labeling
- Recordkeeping
- Permit requirements
- Tracking movement of hazardous wastes.

Today the Department of Transportation (DOT) has new regulations known as HMR-181 (Hazard Materials Regulations - Docket 181). These are contained in 49 CFR (Code of Federal Regulations) parts 100 through 177. Hazardous materials and hazard materials shipments are classified according to hazard classes. **Table II** provides a summary of hazard classes which describe the nature of the material regulated in terms of the type of hazard posed to the shipper and individuals handling the materials, as well as the general public and the environment. Under DOT regulations, hazard classes are further categorized by the severity of the hazard through Packing Group designations. A packing group (PG) denotes the severity of the hazard. There are three Packing Group designations:

PG I (Packing Group I) refers to high hazard; PG II refers to medium danger; PG III refers to minimum danger.

Generators of hazardous wastes are responsible for complying with all DOT legislation regulating DOT and EPA rules covering hazardous wastes. **Table III** lists major areas covered with a summary of requirements, agency responsibility, and CFR reference.

WASTE MINIMIZATION

Waste minimization consists of source reduction and recycling. **Source reduction** is any activity that reduces or eliminates the generation of waste at the source (usually within a process). **Recycling** is recovery

and/or reuse of an otherwise waste material. Incentives for waste minimization include:

- **Economics** - reduction of treatment/disposal costs and savings in raw materials.
- **Liability** reduction.
- **Environmental** concerns.

Waste minimization economics have become more attractive in recent years because of the dramatically increasing costs of waste disposal and their effects by environmental regulations. Careful planning and organization is necessary to bring about a successful **waste minimization** program. An assessment serves to identify the best options for minimizing waste through a thorough understanding of the waste generating processes, waste streams, and operating procedures. Waste generators as well as those who dispose of wastes while striving to meet treatment and disposal rules, realize that reducing waste at the source is a key to their efforts. Too much hazardous waste is forcing industry to look for minimization methods.

In addition to waste streams, information needed to fully understand operations generating wastes include:

- Process, equipment and facility design
- Environmental reports, manifests, documents and permits
- Raw material production information
- Operating costs
- Organization

Table IV provides a listing of accepted hazardous waste treatment technologies. **Table V** lists developing thermal technologies for hazardous waste treatment. **Table VI** lists leachate treatment technologies. These technologies have traditionally been used or considered for "end-of-pipe" treatment of wastes. It now becomes the challenge to incorporate some of these options as an integral part of process design to minimize wastes. The first step in waste minimization management is a thorough auditing of the process waste streams and performing material balances to quantify amounts and concentrations. The subject of material balances is covered later in this text. Finally, **Table VII** lists major contributors of industrial waste streams.

TABLE II.

DEPARTMENT OF TRANSPORTATION HAZARD CLASSES

HAZARD CLASS	DEFINITION
Flammable liquid	Any liquid having a flash point below 100°F.
Combustible liquid	Any liquid having a flash point at or above 100°F and below 200°F.
Flammable solid	Any solid material, other than an explosive, liable to cause fires through friction or retained heat from manufacturing or processing or which can be ignited readily, creating a serious transportation hazard because it burns vigorously and persistently.
Oxidizer	A substance, such as chlorate, permanganate, inorganic peroxide, or a nitrate, that yields oxygen readily to stimulate the combustion of organic matter.
Organic peroxide	An organic compound containing the bivalent -0-0- structure and which may be considered a derivative of hydrogen peroxide where one or more of the hydrogen atoms have been replaced by organic radicals.
Corrosive	Liquid or solid that causes visible destruction or irreversible alterations in human skin tissue at the site of contact, including liquids that severely corrode steel.
Flammable gas	A compressed gas that meets certain flammability requirements.
Nonflammable gas	A compressed gas other than a flammable gas.
Poison A	Extremely dangerous poison gases or liquids belong to this class. Very small amounts of these gases or vapors of these liquids, mixed with air, are dangerous to life.
Poison B	Substances, liquids or solids (including pastes and semisolids), other than poison A or irritating materials, that are known to be toxic to humans. In the absence of adequate data on human toxicity materials are presumed to be toxic to humans if they are toxic to laboratory animals.
Etiologic agents	A viable microorganism, or its toxin, which causes or may cause human disease. These materials are limited to agents listed by the Department of Health and Human Services.

TABLE II.   (continued)

DEPARTMENT OF TRANSPORTATION HAZARD CLASSES

HAZARD CLASS	DEFINITION
Radioactive material	A material that spontaneously emits ionizing radiation having a specific activity greater than 0.002 microcuries per gram (μCi/g).
Explosive	Any chemical compound, mixture, or device, the primary or common purpose of which is to function by explosion, unless such compound, mixture, or device is otherwise classified.
Class A	Detonating explosives.
Class B	Explosives that generally function by rapid combustion rather than detonation.
Class C	Manufactured articles, such as small arms ammunition, that contain restricted quantities of class A and/or class B explosives, and certain types of fireworks.
Blasting agent	A material designed for blasting, but so insensitive that there is very little probability of ignition during transport.
ORM (other regulated materials)	Any material that does not meet the definition of the other hazard classes. ORMs are divided into five substances:
ORM-A	A material which has an anesthetic, irritating, noxious, toxic, or other similar property and can cause extreme annoyance or discomfort to passengers and crew in the event of leakage during transportation.
ORM-B	A material capable of causing significant damage to a transport vehicle or vessel if leaked. This class includes materials that may be corrosive to aluminum.
ORM-C	A material which has other inherent characteristics not described as an ORM-A or ORM-B but which make it unsuitable for shipment unless properly identified and prepared for transportation.
ORM-D	A material such as a consumer commodity which, although otherwise subject to regulation, presents a limited hazard during transportation due to its form, quantity, and packaging.
ORM-E	A material that is not included in any other hazard class but is subject to the requirements of this subject to the requirements of this subchapter. Materials in this class include hazardous wastes and hazardous substances.

TABLE III.

EPA AND DOT HAZARDOUS WASTE TRANSPORTATION REGULATIONS

Required of	Agency	Code of Federal Regulation
Generator/shipper		
• Determine if waste is hazardous according to EPA listing criteria	EPA	40 CFR 261 and 262.11
• Notify EPA and obtain I.D. number; determine that transporter and designated treatment, storage, or disposal facility have I.D. numbers	EPA	40 CFR 262.12
• Identify and classify waste according to DOT Hazardous Materials Table and determine if waste is prohibited from certain modes of transport	DOT	49 CFR 172.101
• Comply with all packaging, marking, and labeling requirements	EPA	40 CFR 262.32(b)
	DOT	49 CFR 173, 49 CFR 172, subpart D, and 49 CFR 172, subpart E 40 CFR 262, subpart D
Transporter/carrier		
• Notify EPA and obtain I.D. number	EPA	40 CFR 263.11
• Verify that shipment is properly identified, packaged, marked, and labeled and is not leaking or damaged	DOT	49 CFR 174-177
• Apply appropriate placards	DOT	49 CFR 172.506
• Comply with all manifest requirements (e.g., sign the manifest, carry the manifest, and obtain signature from next transporter or owner/operator of designated facility)	DOT EPA	49 CFR 174-177 40 CFR 263.20
• Comply with record-keeping and reporting requirements	EPA	50 CFR 263.22
• Take appropriate action (including cleanup) in the event of a discharge and comply with the DOT incident reporting requirements	EPA	40 CFR 263.30-31

TABLE IV.

HAZARDOUS WASTE TREATMENT TECHNOLOGIES

Physical Treatment Methods

- Filtration
- Centrifugation
- Distillation
- Carbon adsorption
- Membrane technology

- moisture removal from solids/sludges
- liquids - solids separation
- solvent separation
- organics removal
- organics - metals removal

Chemical Treatment Methods

- Precipitation
- Oxidation
- Reduction-dechlorination
- Photolysis

- metals
- organics destruction
- chlorine reduction in hydrocarbons
- dioxin/cyanide destruction

Biological Treatment

- Aerobic/Anaerobic
- Land treatment

- metals/organics removal
- organic sludge degradation

Stabilization/Solidification

- Sorption
- Pozzolanic reactions

- immobilization of pollutants
- solidifies inorganic wastes using lime-flyash or Portland cement

TABLE V.

DEVELOPING THERMAL TECHNOLOGIES FOR HAZARDOUS WASTE TREATMENT

Wet Oxidation	For dilute aqueous wastes that cannot be incinerated or biologically treated; destruction efficiencies expected in range of 99% to 99.99% oxidizes organics and inorganics; not suitable for halogenated aromatics.
Vertical-Tube Reactor	Version of wet oxidation into 1-mile deep well system; operates at lower pressures than conventional process. Currently applied to municipal wastewater.
Supercritical Water Oxidation	Oxidizes organics converted to carbon dioxide and water; high pressure steam or electricity produced; inorganic salts precipitated; especially efficient with highly concentrated organic wastes; for water containing 10% organics, destruction efficiency greater than 99.99%. Suitable for chlorinated solvents and PCBs.
Circulating Bed Incineration	High heat-transfer and turbulence allow operation at temperatures lower than traditional incinerators; accommodates solid and liquid wastes; complete destruction of organics at relatively low temperatures; no need for scrubber system to remove acid gases. Is particularly cost-efficient for homogeneous wastes from oil and petro-chemicals.
High-Temperature Fluid-Bed Reactor	Suitable for contaminated soil; liquid wastes require a Wall carrier; pyrolyze organics to carbon, carbon monoxide and hydrogen; equipment not attacked by inorganic components; mobile units possible; reaches high destruction efficiencies.
Pyrolyzing Rotary Reactor	Operates in oxygen-free environment and at lower temperatures than conventional kilns. Produces gas suitable for energy recovery or further treated to recover condensed hydrocarbons; recovery of metals possible without volatilization; reduced need for air pollution control.
Rollins Rotary Reactor	Suitable for viscous and high-solids content wastes; does not require supplemental fuel; reduced gas scrubbing requirements; high-transfer efficiencies increase destruction efficiencies at lower temperatures.
Plasma Arc	Destruction of PCBs and PCB-contaminated equipment; destruction and removal efficiency of 99.9999%; possibly for metal recovery from molten slag.
Pyroplasma Processes	Break-down of waste fluids to elemental constituents; being developed as a mobile unit; for destruction of chlorinated organics; low power consumption and rapid start-stop mode are advantages.

TABLE V. (continued)

DEVELOPING THERMAL TECHNOLOGIES FOR HAZARDOUS WASTE TREATMENT

Plasmadust Process — Metals recovery from iron and steel mill baghouse dusts, reduces metal oxide to elemental forms; iron removed with molten slag; zinc and lead removed as gas; tests show yields of 96% for iron, zinc, and lead.

Penberthy Pyro-Converter — Glass-melting furnace technology adapted for destruction of organics; suitable for liquids, vapors, solids and sludges; solids inorganic residues incorporated into glass matrix. Current use for production of HC1 and destruction of chlorinated organics.

Enzyme Destruction — Biological destruction of organics not involving living organisms; can be maintained in immobilized systems or applied directly to wastes or contaminated material.

UV Photolysis — Used to detoxify liquids containing dioxin, being developed for application on contaminated solids. Dioxin mobilized by surfactants and subjected to UV photolysis; can reduce concentration by 90 to 99%.

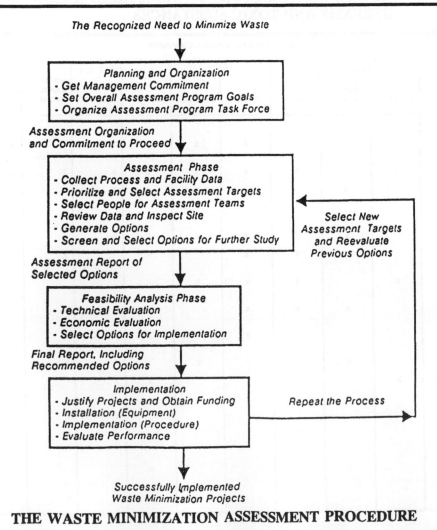

THE WASTE MINIMIZATION ASSESSMENT PROCEDURE

TABLE VI.   LEACHATE TREATMENT OPTIONS

PHYSICAL PROCESSES	CHEMICAL PROCESSES	BIOLOGICAL PROCESSES
Screening	Chemical Precipitation	Aerobic
Mixing	Adsorption	- activated sludge
Sedimentation	Disinfection	- trickling filter
Flotation	Incineration	- aerated lagoon
Elutriation	Aeration	Anaerobic
Vacuum filtration	Oxidation - reduction	- lagoon
Heat transfer	Foam separation	- digestors
Drying	Crystallization	
Reverse osmosis	Ion exchange	
Molecular sieves	Distillation	
Centrifuging		

TABLE VII.   MAJOR CONTRIBUTORS OF INDUSTRIAL HAZARDOUS WASTE STREAMS AND TOXIC SUBSTANCES

INDUSTRY	ARSENIC	CADMIUM	CHLORINATED HYDROCARBONS	CHROMIUM	COPPER	CYANIDE	LEAD	MERCURY	MISCELLANEOUS ORGANICS	SELENIUM	ZINC
MINING & METALLURGY	X	X		X	X	X	X	X		X	X
PAINT & DYE		X		X	X	X	X	X	X	X	
PESTICIDE	X		X			X	X	X	X		X
ELECTRICAL & ELECTRONIC			X		X	X	X	X		X	
PRINTING	X			X	X		X		X	X	
ELECTROPLATING		X		X	X	X					X
CHEMICAL MFG.			X	X	X			X	X		
EXPLOSIVES	X					X	X	X	X		
RUBBER & PLASTICS			X			X		X	X		X
BATTERY		X					X	X			X
PHARMACEUTICALS	X							X	X		
TEXTILE				X	X				X		
PETROLEUM & COAL	X		X				X				
PULP & PAPER								X	X		

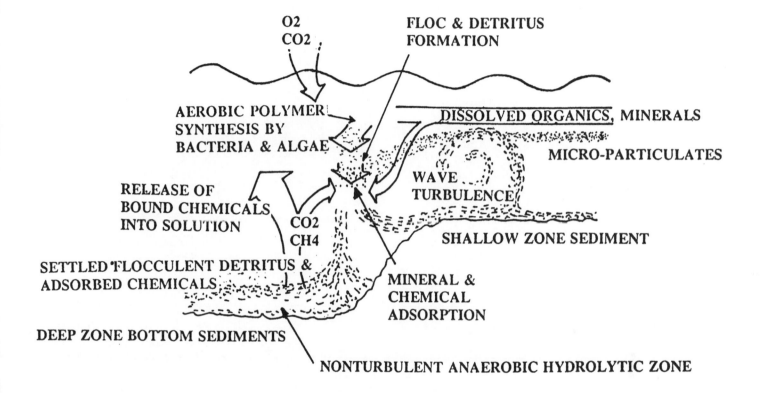

Schematic illustration depicting the interreactions among dissolved organic and inorganic pollutants with inorganic pollutants with suspended microparticles and polymers synthesized by microorganisms.

CHAPTER 2

HAZARDOUS WASTE MANAGEMENT SYSTEMS

OVERVIEW

Technical alternatives in the management of hazardous wastes are:

- volume reduction with disposal
- detoxification
- resource recovery (materials/energy)

Land disposal or burial is becoming less and less of an option because of environmental and regulatory constraints. Volume reduction reduces the magnitude of the problem or if effected by such processes as combustion or thermal treatment methods results in detoxification. Detoxification prior to or during disposal makes the waste innocuous or environmentally adequate. Resource recovery is obviously attractive as a means of reducing hazardous material quantities as well as conserving materials and may offer economic benefits. All of the management options fall individually or in combinations of categories, namely physical, chemical, biological or thermal treatment technologies.

There are usually a range of treatment alternatives for various types of hazardous wastes and their components dependent upon physical and/or chemical properties. Technical alternatives may function similarly in removal or detoxifying certain types of hazardous constituents. However, similarly functioning processes on similar types of hazardous wastes are not necessarily technically equivalent. Differences may occur in allowable inlet feed stream concentrations, or as well as effects of interferences by other waste stream components as well as throughput rates.

Waste sludges are an area of high priority hazardous wastes management. This is especially true with sludges resulting from wastewater treatment since they represent the concentrations of more dilute effluents from a wide variety of sources. Sludges of this nature can contain organics; inorganics; combinations and numerous species placing them in the hazardous materials class. Microorganisms, biologically and for example in activated sludge systems have a long history of degrading chemicals in aqueous waste streams. Application of selected microorganisms to detoxify pollutants in hazardous wastes; contaminated soils; lagoons or groundwater has found great interest in both application and research. Mutation microorganisms show promise for potential in developing new pollution control technology by genetic engineering. Enzymes offer sufficient catalyst activity to employ otherwise slow chemical reactions in pollutant detoxification. This chapter provides an overview of technologies in managing hazardous wastes.

BIOLOGICAL TREATMENT

Biological treatment is applicable to aqueous streams with organic contaminants. The influent waste stream may contain either dissolved or insoluble organics amenable to biodegradation. The microorganisms rely on enzymes for organic decomposition; the enzymes require water to remain active. In aerobic biological treatment, simple and complex organics are eventually decomposed to carbon dioxide and water; oxygen is essential to the hydrocarbon decomposition. In anaerobic biological treatment only simple organics such as carbohydrates, proteins, alcohols and acids can be decomposed.

Biological treatment processes do not destroy or alter inorganics. Concentrations of soluble inorganics should be low so as not to inhibit enzyme activity. Traces of inorganics may be partly removed from liquid waste streams as a result of adsorption on microbial cell coating. Microorganisms typically have a net negative charge and undergo cation exchange with soluble metal ions. Anions such as chlorides or sulfates are not affected by biological treatment.

The principal biological treatment processes are activated sludge, trickling filter, aerated lagoon, enzyme treatment, waste stabilization ponds, anaerobic digestion, and composting. All of these biological systems except enzyme treatment contain living organisms.

Primary sludge - Most wastewater treatment plants use primary sedimentation to remove readily settleable solids from raw wastewater. In a typical plant with primary sedimentation and a conventional activated sludge process for secondary treatment, the dry weight of primary sludge solids is roughly 50 percent of that for the total sludge solids. Primary sludge is usually easier to manage than biological and chemical sludges. Primary sludge is readily thickened by gravity, either within a primary sedimentation tank or within a separate gravity thickener. In comparison with biological and many chemical sludges, primary sludge with low conditioning requirements can be mechanically dewatered rapidly. The dewatering device will produce a drier cake and give better solids capture than it would for most biological and chemical sludges.

Primary sludge production is typically within the range of 800 to 2,500 pounds per million gallons (100 to 300 mg/1) of wastewater. A basic approach to estimating primary sludge production for a particular plant is by computing the quantity of total suspended solids (TSS) entering the primary sedimentation tank and assuming an efficiency of removal. When site-specific data are not available for influent TSS, estimates of 0.15 to 0.24 pound per capita per day (0.07 to 0.11 kg/capita/day) are commonly used. Removal efficiency of TSS in primary sedimentation tanks is usually in the 50 to 65 percent range. An efficiency of 60 percent is frequently used for estimating purposes, subject to the following conditions:

- That the sludge is produced in treatment of a domestic wastewater without major industrial loads.

- That the sludge contains no chemical coagulants or flocculents.

- No other sludges - for example, trickling filter sludge - have been added to the influent wastewater.

- Sludge contains no major sidestreams from sludge processing.

As an example, if a designer estimates the TSS entering the primary clarifier as 0.20 pound per capita per day (0.09 kg/capita/day), and the removal efficiency of the clarifier as 60 percent, the estimated primary sludge production is 0.12 pound per capita per day (0.054 kg/capita/day). If relevant data are available on influent wastewater suspended solids concentrations, such data should, of course, be used for design purposes. Estimates of TSS removal efficiency in primary sedimentation tanks may be refined by the use of operating records from in-service tanks or by laboratory testing. The "Standard Methods" dry weight test for settleable matter estimates under ideal conditions the amount of the sludge produced in an ideal sedimentation tank. Sludge production may be slightly lower in actual sedimentation tanks.

Suspended solids removal efficiency in primary sedimentation depends to a large extent on the nature of the solids. It is difficult to generalize about the effect that industrial suspended solids can have on removal efficiency, but operating experience shows clearly that the amount of sludge withdrawn from the primary sedimentation tank is greatly increased when sludge treatment process sidestreams such as digester supernatant, elutriate and filtrates or concentrates and other sludges like waste-activated sludge are recycled to the primary sedimentation tank. Quantifying the solids entering and leaving the primary clarifier by all streams is an important tool for estimating primary sludge production when recycled sludges and sludge process sidestreams contribute large quantities of solids.

CHEMICAL PRECIPITATION AND COAGULATION

When chemicals are added to the raw wastewater for removal of phosphorus or coagulation to nonsettleable solids, large quantities of chemical precipitates are formed. The quantity of chemical solids produced in chemical treatment of wastewater depends upon the type and amount of chemicals added, chemical constituents in the wastewater, and the performance of the coagulation and clarification processes. It is difficult to predict accurately the quantity of chemicals solids that will be produced. Classical jar tests are favored as a means for estimating chemical sludge quantities. Conditions that influence primary sludge concentration include:

- If the wastewater is not degritted before it enters the sedimentation tanks, the grit may be removed by passing the raw primary sludge through cyclonic separators (called hydroclones). However, these separators do not function properly with sludge concentrations above one percent.

- Industrial loads may strongly affect the primary sludge concentration.

- Primary sludge may float when buoyed up by gas bubbles generated under anaerobic conditions. Conditions favoring gas formation include: warm temperatures; solids deposits within sewers; strong septic wastes; long detention times for wastewater solids in the sedimentation tanks; lack of adequate prechlorination; and recirculating sludge liquors.

- To prevent the septic conditions that favor gas formation, it may be necessary to strictly limit the storage time of sludge in the sedimentation tanks. This is done by increasing the frequency and rate of primary sludge pumping.

- If biological sludges are mixed with the wastewater, a lower primary sludge concentration will generally result.

BIOLOGICAL SLUDGES

Biological sludges are produced by treatment processes such as activated sludge, trickling filters, and rotating biological contactors. Quantities and characteristics of biological sludges vary with the metabolic and growth rates of the various microorganisms present in the sludge. The quantity and quality of sludge produced by the biological process is intermediate between that produced in no-primary systems and that produced in full-primary systems in cases when fine screens or primary sedimentation tanks with high overflow rates are used. Biological sludge containing debris such as grit, plastics, paper and fibers will be produced at plants lacking primary treatment. Plants with primary sedimentation normally produce a fairly pure biological sludge. The concentrations and, therefore, the volumes of waste biological sludge are greatly affected by the method of operation of the clarifiers. Biological sludges are generally more difficult to thicken and dewater than primary sludge and most chemical sludges. The quantity of waste-activated sludge (WAS) is affected by two parameters: the dry weight of the sludge and the concentration of the sludge. The most important variables in predicting waste-activated sludge production are the amounts of organics removed in the process, the mass of micro-organisms in the system, the biologically inert suspended solids in the

influent to the biological process, and the loss of suspended solids to the effluent.

These variable can be assembled into two simple and useful equations:

(1) $P_x = (Y) (s_r) - (k_d) (M)$
(2) $WAS_T = P_x = I_{NV} - E_T$

where:

$P_x =$ net growth of biological solids (expressed as volatile suspended solids (VSS), lb/day or kg/day;

$Y =$ gross yield coefficient, lb/lb or kg/kg;

$s_r =$ substrate (for example, BOD_5) removed, lb/day or kg/day; BOD stands for Biological Oxygen Demand;

$k_d =$ decay coefficient, day^{-1};

$M =$ system inventory of microbial solids (VSS) microorganisms, lb or kg;

$WAS_T =$ waste-activated sludge production, lb/day or kg/day;

$I_{NV} =$ non-volatile suspended solids fed to the process, lb/day or kg/day;

$E_T =$ effluent suspended solids, lb/day or kg/day.

Effect of Feed Composition - The type of wastewater that is fed to the activated sludge process has a major influence on the gross yield (Y) and decay (k_d) coefficients. Many industrial wastes contain large amounts of soluble BOD_5 but small amounts of suspended or colloidal solids. These wastes normally have lower Y coefficients than are obtained with domestic primary effluent. On the other hand, wastes with large amounts of solids, relative to BOD_5, either have higher Y coefficients or require adjustments to reflect the influent inert solids. Even among soluble wastes, different compositions will cause different yields.

Effect of Dissolved Oxygen Concentration - Various dissolved oxygen (DO) levels have been maintained in investigations of activated sludge processes. Very low DO concentrations - for example, 0.5 mg/l - in conventional activated sludge systems do appear to use increased solids production, even when other factors are held constant. However, there is a vigorous disagreement concerning solids production at higher DO levels. Some investigators state that use of pure oxygen instead of air reduces sludge production. This is attributed to the high DO levels attained through the use of pure oxygen. Other investigators have concluded that if at least 2.0 mg/l DO is maintained in air-activated sludge systems, then air and oxygen systems produce the same yield at equivalent conditions (such as food-to-microorganism ratio).

Effect of Temperature - The coefficients Y (gross yield) and k_d (decay) are related to biological activity and, therefore, may vary due to temperature of the wastewater. This variation has not been well documented. One study obtained no significant difference due to temperature over the range 39° to 68°F (4° to 20°C). However, others have observed significant differences within the same temperature range. Sometimes a simple exponential ("Arrhenius") equation is used for temperature corrections to Y and k_d. For instance, it has been stated that

chemical and biochemical rates double with an 18°F (10°C) rise in temperature. Exponential equations have been found to be accurate for pure cultures of bacteria, but are quite inaccurate when applied to Y and k_d for the mixed culture found in real activated sludges. The following guidelines are recommended until such time as process investigations and research efforts in this area provide more consistent and reliable information:

- Wastewater temperatures in the range of from 59° to 72°F (15° to 22°C) may be considered to be a base case. Most of the available data are from this range. Within this range, there is no need to make temperature corrections. Any variations in process coefficients across this temperature range are likely to be small in comparison to uncertainties caused by other factors.

- If wastewater temperatures are in the range of from 50° to 59°F (10° to 15°C), the same k_d value as for 59° to 75°F (15° to 22°C) should be used, but the Y value should be increased by 26 percent. This is based on experiments that compared systems at 52°F (11°C) and 70°F (21°C). In these tests k_d was the same, but Y was 26 percent higher. (On a COD basis, Y was found to be 0.48 at 38°F (11°C) and 0.38 at 4\56°F (21°C).

If wastewater temperatures are below 50°F (10°C), increased sludge production should be expected, but the amount of increase cannot be accurately predicted from available data. Under such conditions, there is a need for pilot-scale process investigations.

If wastewater temperatures are above 72°F (22°C), values of the process coefficients from the range 59° to 72°F (15° to 22°C) may be used for design. The resulting design may be somewhat conservative.

Effective of Feed Pattern - Various feed patterns for the activated sludge process include contact stabilization, step feeding, conventional plug-flow and complete-mix. For design purposes, it appears to be best to ignore the feed pattern when estimating solids production.

Solids Handling - Solids handling facilities must have the capacity to accept wasted solids. For wastewater treatment plants without major known BOD_5 and SS loading variations, allowance should be made in designing solids processing facilities for the wasting of an additional two percent of M per day and lasting up to two weeks. Industrial loads would be either small or unusually stable.

For plants with major variations in loads, allowance should be made for wasting an additional five percent of M per day and lasting for up to two weeks. A similar allowance should be made for plants that practice nitrification during only part of the year. For plants with major weekday-to-weekend variations of over 2 to 1 in BOD_5 load, and medium or high food-to-microorganism ratios of over 0.3 during the high loads, allowance should be made for a one-day sludge wasting of up to 25 percent of M. The plant should also be able to handle wasting of five percent of M per day and lasting for two weeks. Plants in this category serve major industrial systems, large office complexes, schools and ski areas.

Since inventory reductions are not generally practiced during peak loading periods, these capacity allowances should be added to average solids production. The maximum rate of waste-activated sludge production is determined by whichever is greater: production during peak loading or the sum of average production plus inventory reduction allowances. Occasionally, sludge is wasted in a pattern so that M increases at some times and decreases at others. An example of such a pattern is the withdrawal of WAS only during the daytime. Use of such patterns will, of course, increase the maximum rate at which WAS must be removed.

THERMAL TREATMENT/INCINERATION

Methods used to modify hazardous waste previously destined for landfill include incineration and/or some form of thermal treatment. Incinerators and alternate thermal treatment methods are achieving great interest in waste management with considerable efforts into the incinerability of hazardous compounds. Ideally, incineration of wastes to carbon dioxide and inorganic constituents offers a flexible and economical detoxification method for a wide variety of nonrecoverable hazardous wastes. This approach is however not without concern because of the potential concern for emission of toxic combustion by-products such as polychlorinated dioxins and furans.

Total incineration can be defined as the conversion of waste to a solidified residue/slag/or ash and flue gases which consist mainly of carbon dioxide, oxygen, nitrogen and water vapor. While energy recovery may be an option, most systems must include adequate air pollution control. In conventional incineration, temperatures are on the order of 1800°F; total incineration operates at temperatures approaching 3000°F. Incineration objectives include:

- Maximum volume and weight reduction.

- Complete oxidation or combustion of all combustibles with a resultant residue sterile, free of putrescibles, compact, dense and strong, as free of leachable materials as possible.

- A minimal and environmentally acceptable residue disposal operation.

- Complete oxidation of gaseous products of combustion, with an atmospheric discharge after adequate air pollution control.

Incineration/combustion/thermal treatment processes in use and under continued development include:

Liquid Injection Incineration - Limited to destruction of pumpable wastes of viscosities less than 10,000 SS. Usually designed to burn specific waste streams and also used in conjunction with other incinerator systems as a secondary afterburner for combustion of volatiles.

Rotary Kilns - While this technology usually involves a high capital cost, there is an economy of scale. It can accommodate a great variety of waste feeds: solids, sludges, liquids, some bulk waste contained in fiber drums. Combustion chamber rotation enhances waste mixing.

Cement Kilns - Attractive for destruction of harder-to-burn waste, due to very high residence times, good mixing and high temperatures. Alkaline environment neutralizes chlorine.

Boilers - This is usually a liquid injection design that has energy value recovery and fuel conservation. The availability on sites of waste generators reduces spill risks during transportation.

Multiple Hearth - Waste passage onto progressively hotter hearths can provide long residence time for sludges and provides good fuel efficiency. Disadvantages include inability to handle fusible ash, high maintenance due to moving parts in the high temperature zone and cold spots inhibit even and complete combustion.

Fluid-Bed Incinerators - These offer large economy of scale. A turbulent bed enhances uniform heat transfer and combustion of waste. The mass of the bed is relatively large in comparison to the mass of injected waste.

Pyrolysis - Air pollution requirements are minimum; air starved combustion avoids volatilization of inorganic compounds, and these and heavy metals go into insoluble solid char. This process has the potential for high capacity. Disadvantages are that some wastes produce a tar that is hard to dispose of and the method has potentially high fuel and maintenance costs.

Emerging Thermal Processes Include:

Molten Salt - The molten salts act as catalysts and promote heat transfer which reduce energy and maintenance costs. Units are compact and potentially portable. Air pollution control requirements appear minimal since combustible products such as ash and acidic gases are retained in the melt. Regeneration or ash disposal of contaminated salt appears to be a potential problem making the method unsuitable for high ash wastes. Chamber corrosion can be a problem.

Plasma Arc - Very high energy radiation breaks chemical bonds directly without a series of chemical reactions. Operation is simple, low energy costs, and mobile units are feasible. Principal limitations are limited throughput and high NaOH concentrations for scrubbers.

Wet Oxidation - Applicable to aqueous wastes too dilute for incineration and too toxic for biological treatment. Lower temperature requirements and released energy by some wastes are advantages as well as no air emissions. The process is not applicable to highly chlorinated organics and some wastes may require further treatment.

Super Critical Water - Applicable to chlorinated aqueous wastes which are too dilute to incinerate. The process takes advantage of water's excellent solvent properties above the critical point for organic compounds. Injected oxygen decomposes small organic molecules to carbon dioxide and water. There are essentially no air emissions. Economics of scale and energy needs are unclear on scale-up.

High-Temperature Fluid Wall - Wastes can be efficiently destroyed as it passes through the cylinder and is exposed to radiant heat temperatures of about 4000°F. The cylinder is electrically heated; heat is transferred through an inert gas blanket to the waste. Mobile units are possible. To date only limited laboratory and pilot units have been available and scale up remains to be seen. A major disadvantage appears to be the potentially high costs for electrical heating.

Incineration and some of the emerging technologies are receiving increasing attention and a hope for a promise for future waste management. Hard scientific information concerning destruction efficiency and chemical processes in the combustion zone are still desirable. RCRA has affected design of hazardous wastes incinerators by requiring specific destruction and removal efficiencies and has raised questions regarding the incinerability of hazardous compounds. However the wastes' incineration to carbon dioxide and inorganic constituents offers advantageous detoxification methods. A question being raised is the risks and hazards posed by incineration residues and it appears that stabilization of inorganic wastes from such sources may become necessary before landfilling to reduce the environmental impact. Stabilization requires extensive testing procedures to assure that earlier mistakes of land disposal are not repeated.

CONCLUDING REMARKS

Hazardous wastes management needs the consideration of legal requirements determined by political judgements and social values. Public recognition and fear of toxic effects of hazardous chemicals have superimposed additional burdens in the technical requirements for hazardous wastes management. These have been reinforced by serious incidents, health effects and high risk situations, however, the technical expertise exists to solve even the most difficult pollution problems. Disposal of hazardous wastes present varied problems that can be addressed by detoxification procedures.

Table I provides a summary of biological treatment technologies. **Table II** is a summary of aerobic digester design criteria. **Table III** provides data on primary sludge characteristics.

TABLE I.

OVERVIEW OF BIOLOGICAL TREATMENT

Biological Process	Principal Microbial Population	Optimum Temperature	Range in pH	% Solids in Waste Stream	Average Retention Time	Organic Decomposed	Estimated BOD Upper Limits Effectively Handled	Effluent	Residue
Enzyme Treatment	None	Mesophilic	1.5 - 9.5 Varies	<50%	Nil	All can be decomposed by a series of enzymes	No limit	CO_2 & water if complete treatment, otherwise intermediate decomposition products	None
Activated Sludge Treatment	Aerobic Heterotrophic Bacteria	Mesophilic	6 - 8	<1%	<1 Day	All but oil, grease and halogenated aromatics, nitrogen compound	<10,000 mg/l	CO_2 and water 5% - 15% influent BOD remains	Biomass sludge
Trickling Filter	Aerobic Heterotrophic Bacteria	Mesophilic	6 - 8	1%	<1 Day	All but oil, grease and halogenated aromatics, nitrogen	<5,000 mg/l	CO_2 and water 10 - 20% influent BOD remains	Biomass sludge
Aerated-Lagoon	Aerobic Heterotrophic Bacteria and Facultative Anaerobic Heterotrophic Bacteria	Mesophilic	6 - 8	<1%	2-7 Days	All but oil, grease and halogenated aromatic, nitrogen compound	<5,000 mg/l	CO_2 and water 10 - 30% influent BOD remains	Biomass sludge
Water Stabilization Pond	Aerobic Heterotrophic Bacteria and Autotrophic Algae	Mesophilic	6 - 8	<0.1%	3-8 Months	Catophdries proteins, organic acid and alcohol	<100 mg/l	CO_2 and water 10 - 40% influent BOD remains	None
Anaerobic Digestion	Obligate Anaerobic Heterotrophic Bacteria	Thermophilic	6.4 - 7.5	<10%	2 Weeks	Mostly carbohydrate proteins, organic acid and alcohol	Not applicable	Mixed liqpor	Sludge
Composting	Aerobic Heterotrophic Bacteria and Facultative Anaerobic Heterotrophic Bacteria and Algae	Mesophilic / Thermophilic	6 - 8.5	<50%	3 - 8 Months	All organic, phosphorus compounds and nitrogen compounds	No limit	Leachate	None

TABLE II.

SUMMARY OF AEROBIC DIGESTER DESIGN CRITERIA

	Days	Liquid temperature
Solids residence time required to achieve		
40 percent volatile solids reduction	108	40°F
	31	60°F
	18	80°F
55 percent volatile solids reduction	386	40°F
	109	60°F
	64	80°F
Oxygen requirements	2.0 pounds of oxygen per pound or volatile solids destroyed when liquid temperature 113°F or less	
	1.45 pounds of oxygen per pound of volatile solids destroyed when liquid temperature greater than 113°F	
Oxygen residual	1.0 mg/l of oxygen at worst design conditions	
Expected maximum solids concentration achievable with decanting	2.5 to 3.5 percent solids when dealing with a degritted sludge or one in which no chemicals have been added	
Mixing horsepower	Function of tank geometry and type of aeration equipment utilized. Should consult equipment manufacturer. Historical values have ranged from 0.5 to 4.0 horsepower per 1,000 cubic feet of tank volume	

TABLE III.

PRIMARY SLUDGE CHARACTERISTICS

Characteristic	Range of values		Typical value	Comments
pH	5	- 8	6	-
Volatile acids, mg/l as acetic acid	200	- 2,000	500	-
Heating Value, Btu/lb (kJ/kgk)	6,800	- 10,000	-	Depends upon volatile content, and sludge composition, reported values are on a dry weight basis.
			10,285	Sludge 74 percent volatile.
			7,600	Sludge 65 percent volatile.
Specific gravity of individual solid particles	-		1.4	Increases with increased grit, silt, etc.
Bulk specific gravity (wet)	-		1.02	Increases with sludge thickness and with specific gravity of solids.
			1.07	Strong sewage from a system of combined storm and sanitary sewers.
BOD_5/VSS ratio	0.5	- 1.1	-	-
COD/VSS ratio	1.2	- 1.6	-	-
Organic N/VSS ratio	0.05	- 0.06	-	-
Volatile content, percent by weight of dry solids	64	- 93	77	Value obtained with no sludge recycle, good degritting, 42 samples, standard deviation 5.
	60	- 80	65	
	-		40	Low value caused by severe storm inflow.
	-		40	Low value caused by industrial waste.
Cellulose, percent by weight of dry solids	8	- 15	10	-
	-		3.8	-
Hemicellulose, percent by weight of dry solids	-		3.2	-
Lignin, percent by weight of dry solids	-		5.8	-
Grease and fat, percent by weight of dry solids	6	- 30	-	Ether soluble
	7	- 35	-	Ether extract

TABLE III.   (continued)

PRIMARY SLUDGE CHARACTERISTICS

Characteristic	Range of values	Typical value	Comments
Protein, percent by weight of dry solids	20 - 30 22 - 28	25 -	-
Nitrogen, percent by weight of dry solids	1.5 - 4	2.5	Expressed as N
Phosphorus, percent by weight of dry solids	0.8 - 2.8	1.6	Expressed as P_2O_5. Divide values as P_2O_5 by 2.29 to obtain values as P.
Potash, percent by weight of	0 - 1	0.4	Expressed as K_2O. Divide values as K_2O by 1.20 to obtain values as K.

1 Btu/lb = 2.32 kJ/kg

CHAPTER 3

HAZARDOUS WASTE TREATMENT AND RECOVERY SYSTEMS

INTRODUCTION AND OVERVIEW

New strategies are being sought and undertaken for hazardous wastes treatment and disposal in the face of large scale generation and increasingly stringent regulatory requirements. Hazardous wastes do not exist in isolation and it is increasingly important as we deal with hazardous and solid wastes and wastewaters to understand where they come from and where they go. Additionally important are the possible exposure routes of hazardous wastes or toxic substances by our emissions, pesticide applications and "non-point" waste sources as well as wastes discharged to surface waters through a NPDES (National Pollutant Discharge Elimination System) permit.

Besides the very nature of specific wastes other factors that must be foreseen are future waste volumes; economic and population growth; regulatory decisions and waste minimization efforts. As industrial production increases the volume and variety of wastes could also increase. Recycling and reuse efforts as well as minimization of waste amounts generated can temper waste volume increases. Because of the large number of new statutory and other requirements, hazardous wastes management must deal with "moving targets."

Regulatory impacts, waste disposal costs, liability risks, regional capacity limitations and waste minimization efforts will impact on waste management. Overall factors affecting management practices include:

- Restrictions on land disposal

- Hazardous waste treatment and disposal costs will steadily increase.

- New and more stringent regulations.

WASTE MINIMIZATION

Costs of treating and disposal of hazardous wastes have been and will increase significantly. Rising costs and liability concerns has resulted in the study of ways to minimize the amounts of wastes produced by many companies. According to a recent EPA study, it may be possible to achieve a 20-30 percent reduction in waste through process changes, product substitution, good house-keeping and recycling practices.

Waste minimization activities can include:

- Recycling/reuse on-site.

- Recycling/reuse off-site.

- Equipment/technology modifications.

- Reformulation/redesign of product.

- Raw materials substitution.

- Improved housekeeping, tighter inventory control.

- Discontinue products.

- Source reduction.

- Waste separation and concentration.

- Waste exchange.

Reasons for **waste minimization** action can be described as:

- Regulatory requirements for the waste.

- Reduction of treatment/disposal costs.

- Other process cost reductions.

- Occupational safety.

There are three statutory requirements related to waste minimization enacted in the 1984 HSWA:

- Generators must certify on their manifests. They must have a program in place for volume and toxicity waste reduction (Section 3002(b)).

- Any new treatment, storage or disposal permit must include a waste minimization certification statement (Section 3005(h)).

- As part of the generator's biennial report, generators must describe efforts undertaken during the year to reduce volume and toxicity of waste generated (Section 3002 (a)(b)) and document actual reduction achieved.

The concept of waste minimization incorporates recovery and recycling opera-

tions, waste exchange programs and source reduction techniques. It is far better to reduce the generation of hazardous waste than to manage waste after it is created.

The Hazardous and Sold Waste Amendments of 1984 (HSWA) has banned land disposal of over 400 chemicals and waste streams unless they are treated or demonstrated there will be no migration as long as the waste remains hazardous. The major impact of these rules is to significantly increase treatment requirements as well as add incentives for waste minimization. Under the **Superfund Amendments and Reauthorization Act of 1986** (SARA), EPA is required to establish standards for **Superfund** clean-up actions and identify conditions for disposing of **Superfund** wastes off-site. The impact of these provisions is changing the amounts of hazardous wastes managed on-site and off-site. Mobile treatment units and stabilization methods should ideally be employed for on-site wastes while more concentrated wastes are likely to shift off-site for treatment and disposal. About 96 percent of RCRA hazardous waste is managed on the sites of private companies. The remaining four percent goes to off-site commercial treatment and disposal facilities in current practice.

TREATMENT TECHNOLOGIES FOR WASTE MINIMIZATION

Treatment technologies are designed to change the character of hazardous wastes to render them less hazardous or environmentally acceptable. The most widely used treatment methods include incineration, biological and chemical wastewater treatment, steam stripping and solidification.

Most of the 275 million tons per year of hazardous waste is treated in impoundments and wastewater treatment plants. Incineration accounts for approximately two million tons per year presently, but this volume may increase substantially as a result of the HSWA requirements.

Incineration is used primarily to burn liquid organic hazardous wastes. Additionally some incinerators are designed to burn sludges and solids as well as liquid wastes. Management decisions because of ongoing and future regulations will favor this practice over land disposal for certain wastes. Incineration/combustion is a destruction process and the main recovery potential in this technique is heat or energy recovery if justified by economics and scale of operation. Reuse as fuel involves use of combustible organic wastes as substitutes or supplements for conventional fuels burned in industrial boilers.

Biological and chemical wastewater treatment is the most widely employed method for treating aqueous hazardous waste. Biological decomposition, chemical neutralization, precipitation render wastewaters less hazardous. Retention time in treatment units can vary from a few hours to days depending on temperatures, microorganisms and waste stream types. The residual sludge produced in this treatment process is generally incinerated, further treated or land disposed.

Steam stripping technologies are used in treating aqueous, hazardous wastewaters. Hazardous constituents (volatiles, solvents, etc.) converted to gas or vapor by this physical treatment may be recovered by air pollution control equipment or condensation.

The physical and chemical characteristics of the waste will determine the treatment and disposal practices as well as recovery po-

tential. RCRA hazardous wastes vary from dilute wastewater to metal bearing sludges to PCB contaminated soils. Over 90 percent of RCRA hazardous wastes are in the form of wastewater, the remainder are organic/inorganic sludges and solids.

UNIT OPERATIONS AND RECOVERY PROCESSES

Adsorption is a process in which the pollutant is adsorbed onto the surface of the adsorbent until its capacity is reached. Widely used adsorbent materials are activated carbon, resins and molecular sieve materials. The adsorbent can then be regenerated and the pollutant is released in a more concentrated form that is recovered or treated by further processing. The specific adsorption/regeneration process and the pollutant and process conditions determine further process steps. Further processing steps can include incineration or condensation and decantation so that the chemical can be recovered for recycling or disposal. Adsorption is particularly effective in removal of various toxic chemicals such as volatile organic compounds (VOCs) from air. Typically adsorption capacity increases with the molecular weight of the VOC being adsorbed. Unsaturated compounds are generally more completely adsorbed than saturated compounds, and cyclic compounds are more readily adsorbed than linear structure organics. VOC's characterized by low vapor pressures are more easily adsorbed than those with high vapor pressures and adsorption capacity is enhanced by lower operating temperatures and higher concentrations.

Condensation is a control technique for some organic compounds. This process cools the gas stream thereby transforming the gaseous compound to a liquid. Condensation is the primary method for product recovery as well as an air pollution control technique. Control of storage and process emissions is a common application employing condensers in series with other control equipment (e.g. absorbers, incinerators, adsorbers).

Absorption is a physical or chemical process that transfers components from a gas stream to a liquid and is used to recover products, raw materials and as an emission control device. Absorption recovery examples are alcohols, acids, halogenated compounds, aromatics, esters and aldehydes. Liquid to gas ratios, liquid temperature, column height are some important parameters affecting efficiencies.

Phase separation is employed on waste streams such as slurries, sludges and emulsions which are not single phase. Phase separation allows significant volume reduction, especially if the hazardous component is present in significant amounts in only one of the phases. In concentrating the hazardous portion of a waste stream follow-up processing may be effected more efficiently and economically. Many phase separation processes are mechanical, simple and inexpensive and apply to a broad range of waste components and wastes.

The simplest phase separation process is **sedimentation** or **gravity settling** where the output consists of a sludge and a decantable supernatant liquid. Other phase separation processes for solids concentration are filtration and centrifugation. Flotation is a recovery process for materials from aqueous solutions having a specific gravity of less than one. High gradient magnetic separation is a newer process that may have potential for separating magnetic and paramagnetic particles from slurries. Colloidal slurries

can be separated by such processes as flocculation and ultrafiltration.

Sludge handling phase separation dewatering is effected by vacuum or press filtration as the most widely used techniques. Sludges and slurries in which the liquid phase is volatile may be treated by either evaporation or distillation. Solar evaporation is commonly employed, while engineered evaporation and/or distillation systems are employed if recovery of the volatile liquid is desired.

RECYCLING/REUSE/FUEL: POTENTIAL AREAS AND INCENTIVES

Most wastes currently going to land disposal may require alternate disposal techniques. The most likely alternative widely discussed at present is incineration. Potentials in this latter technology exist in possibilities of energy recovery. Incineration as an alternative at present offers limited excess capacity, particularly for certain liquid organic wastes while on-site capacity is uncertain. Siting and permitting delays can slow capacity expansion. More sludges and solids could be brought into the hazardous waste system through regulatory action.

Large volumes of hazardous wastes are currently treated and disposed of in surface impoundments. A trend towards tank treatment might occur in a shift from surface impoundments. Solvent wastes scheduled for bans have been granted a two year variance if they require tank treatment.

Many inorganic solids and sludges can be potentially solidified. A limiting factor for this option is the availability of landfills.

The advantage of **waste minimization** in many cases is that major permit modifications are not necessary. Fuel specifications and air emissions may have to be considered. EPA's wastes minimization program has two main objectives:

- To foster the use of waste minimization through technology and information dissemination.

- To report to Congress on the need for **minimization regulation**.

Concern over economic and liability issues are the impetus to reduce volume and toxicity of hazardous waste produced. **Waste minimization** can alleviate capacity problems by reducing the volume of waste requiring treatment and disposal.

The reader should review the tables and flowsheets in the remainder of this chapter for specific concepts, technologies, and process options for waste minimization.

Estimate of Physical Characteristics of RCRA Hazardous Wastes

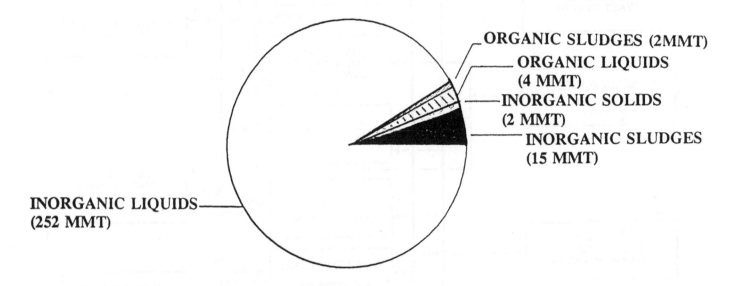

ORGANIC SLUDGES (2MMT)
ORGANIC LIQUIDS (4 MMT)
INORGANIC SOLIDS (2 MMT)
INORGANIC SLUDGES (15 MMT)
INORGANIC LIQUIDS (252 MMT)

TREATMENT TECHNOLOGIES

Physical Characteristics	Incineration	Wastewater Treatment	Treatment Impoundments	Solidification	Steam Stripping
Organic Liquids (4 MMT)	●	○	○		
Inorganic Liquids (252 MMT)	○	●	●	○	○
Organic Sludges (2 MMT)	○	○	○	○	
Inorganic Sludges (15 MMT)	○	○	○	●	
Inorganic Solids (2 MMT)	○			○	

● Widely Used
○ Sometimes Used

Source: EPA, Office of Solid Waste

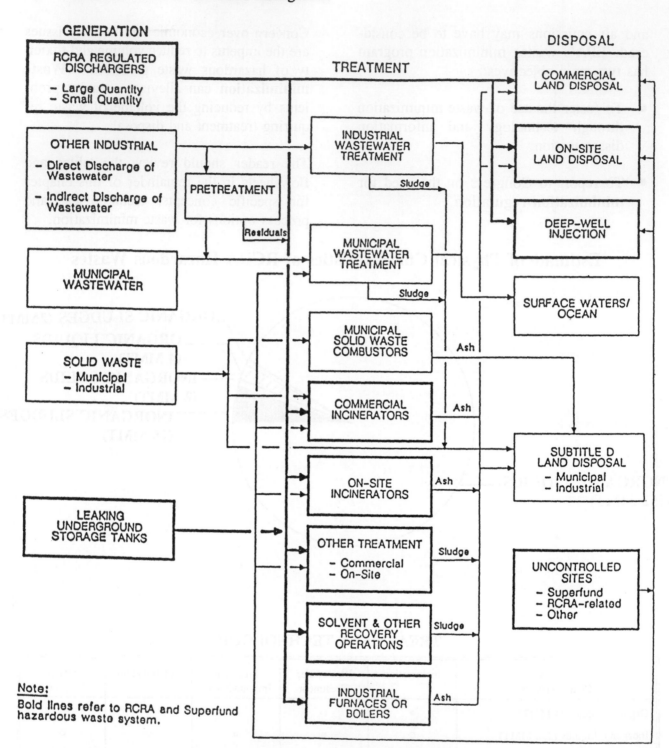

GENERATION

RCRA REGULATED
DISCHARGERS

- Large Quantity
- Small Quantity

OTHER INDUSTRIAL

- Direct Discharge of
  Wastewater
- Indirect Discharge of
  Wastewater

MUNICIPAL
WASTEWATER

SOLID WASTE
- Municipal
- Industrial

LEAKING
UNDERGROUND
STORAGE TANKS

TREATMENT

PRETREATMENT

Residuals

INDUSTRIAL
WASTEWATER
TREATMENT

Sludge

MUNICIPAL
WASTEWATER
TREATMENT

Sludge

MUNICIPAL
SOLID WASTE
COMBUSTORS

Ash

COMMERCIAL
INCINERATORS

Ash

ON-SITE
INCINERATORS

Ash

OTHER TREATMENT

- Commercial
- On-Site

Sludge

SOLVENT & OTHER
RECOVERY
OPERATIONS

Sludge

INDUSTRIAL
FURNACES OR
BOILERS

Ash

DISPOSAL

COMMERCIAL
LAND DISPOSAL

ON-SITE
LAND DISPOSAL

DEEP-WELL
INJECTION

SURFACE WATERS/
OCEAN

SUBTITLE D
LAND DISPOSAL

- Municipal
- Industrial

UNCONTROLLED
SITES

- Superfund
- RCRA-related
- Other

Note:
Bold lines refer to RCRA and Superfund
hazardous waste system.

WASTE SYSTEM CHART

HAZARDOUS WASTE MANAGED BY INDUSTRY

INDUSTRY	NUMBER OF FACILITIES	AMOUNT WASTE MANAGED (MMT)	GENERAL WASTE DESCRIPTION
Chemical	700	218	Contaminated wastewaters, spent solvent residuals, still bottoms, spent catalysts, treatment sludges, and filter cakes
Fabricated Metals	200	4	Electroplating wastes, sludges contaminated with metals and cyanides, degreasing solvents
Electrical Equipment	240	1	Deegreasing solvents
Petroleum Refinery	100	20	Leaded Tank bottoms, slop oil emulsion solids, API separator sludge, DAF float
Primary Metals	150	4	Pickle liquor, sludge with metal contaminates
Transportation Equipment	150	3	Degreasing solvents, metals, sludges
National Security	100	1	All types of wastes
Other	1360	24	All types of wastes
Total	3000	275	

Source: National Screening Survey of Hazardous Waste Treatment, Storage, Disposal and Recycling Facilities, U.S. EPA, Office of Solid Waste, Office of Policy, Planning and Information, 1986.

SOURCES AND TYPES OF INDUSTRIAL WASTES

SIC Group Classification	Waste Generating Processes	Expected Specific Wastes
Plumbing, heating, air conditioning Special trade contractors	Manufacturing and installation in homes, buildings, factories	Scrap metal from piping and duct work; rubber, paper, insulating materials, misc. construction, demolition debris
Ordnance and accessories	Manufacturing and assembling	Metals, plastic, rubber, paper, wood, cloth, chemical residues
Food and kindred products	Processing, packaging, shipping	Meats, fats, oils, bones, offal vegetables, fruits, nuts and shells, cereals
Textile mill products	Weaving, processing, dyeing, and shipping	Cloth and fiber residues
Apparel and other finished products	Cutting, sewing, sizing, pressing	Cloth, fibers, metals, plastics, rubber
Lumber and wood products	Sawmills, mill work plants, wooden container, misc. wood products, manufacturing	Scrap wood, shavings, sawdust; in some instances metals, plastics, fibers, glues, sealers, paints, solvents
Furniture, wood	Manufacture of household and office furniture, partitions, office and store fixtures, mattresses	Those listed under Code 24; in addition, cloth and padding residues
Furniture, metal	Manufacture of household and office furniture, lockers, bedsprings, frames	Metals, plastics, resins, glass, wood, rubber, adhesives, cloth, paper
Paper and allied products	Paper manufacture, conversion of paper and paperboard, manufacture of paperboard boxes and containers	Paper and fiber residues, chemicals, paper coatings and fillers, inks, glues, fasteners
Printing and publishing	Newspaper publishing, printing, lithography, engraving, and bookbinding	Paper, newsprint, cardboard, metals, chemicals, cloth, inks, glues
Chemicals and related products	Manufacture and preparation of inorganic chemicals (ranges from drugs and soups to paints and varnishes, and explosives)	Organic and inorganic chemicals, metals, plastics, rubber, glass, oils, paints, solvents, pigments
Petroleum refining and related industries	Manufacture of paving and roofing materials	Asphalt and tars, felts, asbestos, paper, cloth, fiber
Rubber and miscellaneous plastic products	Manufacture of fabricated rubber and plastic products	Scrap rubber and plastics, lampblack, curing compounds, dyes
Leather and leather products	Leather tanning and finishing; manufacture of leather belting and packing	Scrap leather, thread, dyes, oils, processing and curing compounds

SOURCES AND TYPES OF INDUSTRIAL WASTES

SIC Group Classification	Waste Generating Processes	Expected Specific Wastes
Stone, clay, and glass products	Manufacture of flat glass, fabrication or forming of glass; manufacture of concrete, gypsum, and plaster products; forming and processing of stone and stone products, abrasives, asbestos, and misc. nonmineral products	Glass, cement, clay, ceramics, gypsum, asbestos, stone, paper, abrasives
Primary metal industries	Melting, casting, forging, drawing, rolling, forming, extruding operations	Ferrous and nonferrous metals scrap, slag, sand, cores, patterns, bonding agents
Fabricated metal products	Manufacture of metal cans, hand tools, general hardware, nonelectric heating apparatus, plumbing fixtures, fabricated structural products, wire, farm machinery and equipment, coating and engraving of metal	Metals, ceramics, sand, slag, scale, coatings, solvents, lubricants, pickling liquors
Machinery (except electrical)	Manufacture of equipment for construction, mining, elevators, moving stairways, conveyors, industrial trucks, trailers, stackers, machine tools, etc.	Slag, sand, cores, metal scrap, wood, plastics, resins, rubber, cloth, paints, solvents, petroleum products
Electrical	Manufacture of electric equipment, appliances, and communication apparatus, machining, drawing, forming, welding, stamping, winding, painting, plating, baking, firing operations	Metal scrap, carbon, glass, exotic metals, rubber, plastics, resins, fibers, cloth residues
Transportation equipment	Manufacture of motor vehicles truck and bus bodies, motor vehicle parts and accessories, aircraft and parts, ship and boat building and repairing motorcycles and bicycles and parts, etc.	Metal scrap, glass, fiber, wood, rubber, plastics, cloth, paints, solvents, petroleum products
Professional, scientific controlling instruments	Manufacture of engineering, laboratory, and research instruments and associated equipment	Metals, plastics, resins, glass, wood, rubber, fibers, abrasives
Miscellaneous manufacturing	Manufacture of jewelry, silverware, plated ware, toys, amusement, sporting and athletic goods, costume novelties, buttons, brooms, brushes, signs, advertising displays	Metals, glass, plastics, resins, leather, rubber, composition, bone, cloth, straw, adhesives, paints, solvents

COMPARISON OF TREATMENT PROCESSES THAT SEPARATE
ORGANICS FROM LIQUID WASTE STREAMS

TREATMENT PROCESS	REQUIRED FEED STREAM PROPERTIES	CHARACTERISTICS OF OUTPUT STREAM(S)
Carbon Adsorption	Aqueous solutions; concentrations < 1%; SS < 50 ppm	Adsorbate on carbon; usually regenerated at thermally or chemically
Resin Adsorption	Aqueous solutions; concentrations < 8%; SS < 50 ppm; no oxidants	Adsorbate on resin; always chemically regenerated
Ultrafiltration	Solution or colloidal suspension of high molecular weight organics	One concentrated in high molecular weight organics; one containing dissolved ions
Air Stripping	Solution continuing ammonia; high pH	Ammonia vapor in air
Steam Stripping	Aqueous solutions of volatile organics	Concentrated aqueous streams with volatile organics and dilute stream with residuals
Solvent Extraction	Aqueous or non-aqueous solutions; concentration < 10%	Concentrated solution of organics in extraction solvent
Distillation	Aqueous or non-aqueous solutions; high organic concentrations	Recovered solvent; still bottom liquids, sludges and tars
Steam Distillation	Volatile organics, non-reactive with water or steam	Recovered volatiles plus condensed steam with traces of volatiles

PROCESSES THAT DESTROY ORGANICS

PROCESS	REQUIRED FEED STREAM PROPERTIES	CHARACTERISTICS OF OUTPUT STREAM(S)
Biodegradation	Dilute aqueous streams with soluble organics	Pure water
Oxidation	Dilute aqueous solutions of phenols, organic sulfur compounds, chlorinated hydrocarbons, etc.	Oxidation products in aqueous solutions
Ozonation	Aqueous solutions; concentrations < 1%	Oxidation products in aqueous solutions
Calcination	Organics and inorganics that decompose thermally	Solid oxides; volatile emissions
Hydrolysis	Aqueous or non-aqueous streams; no concentration limits	Hydrolysis products
Photolysis	Aqueous streams; transparent to light; components that absorb radiation	Photolysis products
Chlorinolysis	Chlorinated hydrocarbon waste streams; low sulfur, low oxygen; can contain benzene and other aromatics; no solids; no tars	Carbon tetrachloride; HCl and phosgene HCl and phosgene
Microwave Discharge	Organic liquids or vapors	Discharge products; not accurately predictable

COMPARISON OF PROCESSES THAT SEPARATE
HEAVY METALS FROM LIQUID WASTE STREAMS

PROCESS	REQUIRED FEED STREAM PROPERTIES	CHARACTERISTICS OF OUTPUT STREAM(S)
Physical Removal		
Ion Exchange	Con. <4000 ppm; aqueous solutions, low SS	One concentrated in heavy metals; one purified
Reverse Osmosis	Con. <4000 ppm; aqueous solutions; controlled pH; low SS; no strong oxidants	One concentrated in heavy metals; one with heavy metal concentrations >5 ppm
Electrodialysis	Aqueous solutions; neutral or slightly acidic; Fe and MN <0.3 ppm; CU<400 ppm	One with 1000-5000 ppm heavy metals; one with 100-500 ppm heavy metals
Liquid Ion Exchange	Aqueous solutions; no concentration limits; no surfactants; SS<0.1%	Extraction solvent concentrated in heavy metals; purified water or slurry
Freeze Crystallization	Aqueous solutions; TDS<10%	Concentrated brine or sludge; purified water, TDS ~ 100 ppm
Chemical Removal		
Precipitation	Aqueous or low viscosity non-aqueous solutions; no concentration limits	Precipitated heavy metal sulfides, hydroxides, oxides, etc.; solvent with TDS governed by solubility product of precipitates
Reduction	Aqueous solutions; concentrations of heavy metals <1%; controlled pH	Acidic solutions with reagent (oxidized $NaBH_4$ or Zn); metallic precipitates
Electrolysis	Aqueous solutions; heavy metal concentrations <10%	Recovered metals; solution with 2-10 ppm heavy metals

COMPARISON OF PROCESSES THAT SEPARATE TOXIC ANIONS
FROM LIQUID WASTE STREAMS

PROCESS	REQUIRED FEED STREAM PROPERTIES	CHARACTERISTICS OF OUTPUT STREAM(S)
Physical Removal		
Ion Exchange	Inorganic or organic anions in aqueous solution	Concentrated aqueous solutions
Liquid Ion Exchange	Inorganic or organic anions in aqueous solution	Concentrated solutions in extraction solvent
Electrodialysis	Aqueous stream with 1000-5000 ppm inorganic salts; and pH; Fe and Mn <0.3 ppm	Concentrated aqueous stream (10,000 ppm salts); dilute stream (100-500 ppm salts)
Reverse Osmosis	Aqueous solutions with up to 34,000 ppm total dissolved solids	Dilute solution (~ 5 ppm TDS); concentrated solution of hazardous components
Freeze Crystallization	Aqueous salt solutions	Purified water; concentrated brine
Chemical Removal		
Oxidation	Aqueous solutions of cyanides, sulfides, sulfites etc.; concentrations $< 1\%$	Oxidation products
Ozonation	Aqueous solutions of cyanides; concentrations $< 1\%$	Cyanate solutions
Electrolysis	Alkaline aqueous solutions of cyanides or concentrated HCl solutions ($>20\%$)	Cyanides to ammonium and carbonate salt solutions; HCl to Cl_2 gas

COMPARISON OF PROCESSES THAT CAN ACCEPT SLURRIES OR SLUDGES

PROCESS	REQUIRED FEED STREAM PROPERTIES	CHARACTERISTICS OF OUTPUT STREAM(S)
Calcination	Waste stream with components that decompose by volatilization (hydroxides, carbonates, nitrates, sulfites, sulfates)	Solid greatly reduced in volume; volatiles
Freeze Crystallization	Low-viscosity aqueous slurry or sludge	Brine sludge; purified water
HGMS	Magnetic or paramagnetic particles in slurry	Particles
Liquid Ion Exchange	Solvent extractable inorganic component	Solution in extraction solvent
Flotation	Floatable particles in slurry	Froth
Hydrolysis	Hydrolyzable component	Hydrolysis products
Anaerobic Digestion	Aqueous slurry; < 7% solids; no oils or greases; no aromatics or long chain hydrocarbons	Sludges; methane and CO_2
Composting	Aqueous sludge; < 50% solids	Sludge; leachate
Steam Distillation	Sludge or slurry with volatile organics	Volatile, solid residue
Solvent Extraction	Solvent extractable organic	Solution of extracted components; residual sludge

COMMON AIR RELEASE SOURCE CATEGORIES

Process Vents	Secondary Sources	Fugitive Sources	Handling, Storage, and Loading
Reactors	Pond evaporation	Flanges/connectors	Breathing losses
Distillation system	Cooling tower evaporation	Valves	Filling losses
Vacuum systems	Wastewater treatment	Pump seals	Loading/unloading
Baghouses or precipitators	facilities	Compressor seals	Line venting
Combustion stacks		Sample connections	Packaging/container
Blow molding		Open-ended lines	loading
Spray drying		Pressure relief devices	
Curing/drying		(e.g., rupture disks)	
Scrubbers/absorbers		Lab hoods	
Centrifuges		Process sampling	
Extrusion operations		Equipment inspection	
Pressure safety valves		Equipment cleaning	
Manual ventings		Equipment maintenance	
		Blowing out pipelines	
		Storage piles	

NOTE: Process vents usually point sources.

Secondary sources are usually not contained and are considered fugitive sources.

Storage tank emissions are considered as point sources; other loading and unloading releases could be categorized as either point or fugitive sources, depending on whether the releases are ducted.

TYPICAL WASTEWATER SOURCES

Untreated process wastewater

Miscellaneous untreated wastewater - equipment washdown, steam jet condensate, cooling water

Decantates or filtrates

Cleaning wastes

Steam stripping wastes

Acid leaching solutions

Spent plating, stripping, or cleaning baths

Spent scrubber, absorber, or quench liquid

Off-spec, discarded products or feedstock

Distillation side cuts

Cyclone or centrifuge wastes

Spills, leaks, vessel overflows

METHODS OF WASTEWATER DISPOSAL

Direct discharge to surface waters

Discharge to a publicly owned treatment works

Underground injection

Surface impoundments

Land treatment

REFUSE HEATING VALUES

	Moisture	BTU as fired
Domestic Refuse		
Paper, cardboard, cartons, bags	3	7,660
Wood crates, boxes, scrap	7	7,825
Brush, branches	17	7,140
Leaves	30	4,900
Grass	50	3,820
Garbage	75	1,820
Greenstuff	50	3,470
Greens	50	4,070
Rags, cotton, linen	10	6,440
Industrial Scrap Refuse		
Boot, shoe trim and scrap		8,500
Sponge waffle and scrap		8,500
Butyl soles scrap		11,500
Cement wet scrap		11,500
Rubber		12,420
Tire cord scrap		12,400
Rubber scorched scrap		19,700
Tires, bus and auto		18,000
Gum scrap		19,700
Latex coagulum		19,700
Latex waste, coagulum waste		12,000
Leather scrap		10,000
Waxed paper		12,000
Cork scrap		12,400
Paraffin		16,803
Oil waste, fuel oil residue		18,000
Plastic and Synthetic Refuse		
Cellophane plastic		12,000
Polethylene		19,840
Polyvinyl chloride		17,500
Vinyl scrap		17,500
Aldehyde sludge		18,150
Solvent naptha		18,500
Carbon disulfite		8,000
Benzine		10,000
Miscellaneous		
Carbon to CO_2		14,093
Carbon to CO		4,347
Sulfur		3,983
Methane		23,879

UNIT OPERATIONS AND TREATMENT PROCESSES USED TO TREAT SOLID, SLURRY AND NONAQUEOUS WASTES

Incineration/thermal treatment
    Liquid injection incineration
    Rotary kiln incineration
    Fluidized bed incineration
    Multiple hearth chamber incineration
    Pyrolytic destruction
    Other incineration/thermal treatment

Reuse as fuel
    Cement kiln
    Aggregate kiln
    Asphalt kiln
    Other kiln
    Blast furnace
    Sulfur recovery furnace
    Smelting, melting, and refining furnace
    Coke oven
    Other furnace
    Industrial boiler
    Utility boiler
    Other reuse as fuel unit

    Fuel blending

Solidification
    Cement-based processes
    Pozzolanic processes
    Asphaltic processes
    Thermoplastic techniques
    Organic polymer techniques
    Macro-encapsulation
    Other solidification

Recovery of solvents and other organic chemicals
    Fractionation
    Batch still distillation
    Solvent extraction
    Thin film evaporation
    Other solvent recovery

Recovery of metals
    Activated carbon (for metals recovery)
    Electrodialysis (for metals recovery)
    Electrolytic metal recovery
    Ion exchange (for metals recovery)
    Reverse osmosis (for metals recovery)
    Solvent extraction (for metals recovery)
    Ultrafiltration (for metals recovery)
    Other metals recovery

Dewatering operations
    Gravity thickening
    Vacuum filtration
    Pressure filtration (belt, plate and frame, leaf)
    Centrifuge
    Other dewatering

APPLICATIONS AND EFFICIENCY OF CARBON ADSORPTION

Compound	Control efficiency, %	Comments
Acetone/phenol	92	Overall hydrocarbon removal efficiency
	83.4	Efficiency calculated from design data
	99	Efficiency including condenser
Dimethyl terephthalate	80	VOC removal efficiency
	97	p-Xylene removal efficiency
Maleic anhydride	85	System control efficiency
Methylene chloride	>90	Reported efficiency for controlling emission from pharmaceutical manufacturing
Perchloroethylene	96, 99	Perchloroethylene control efficiency
	97	Test data from dry cleaning industry

APPLICATIONS AND EFFICIENCY OF WET SCRUBBERS

Compound or metal	Accompanying control device	Control efficiency, [a]%	Source of emissions
Adipic Acid		98	SOCMI
Cadmium		98	Primary lead smelting
Manganese		90+	Ferroalloy production
		99+	Iron and steel foundries
		98	Steel production
Phthalic anhydride	Thermal incinerator[b]	96.5[c,d]	SOCMI
Maleic Acid	Maleic acid recovery[e]		
Zinc	Thermal incinerator[b]	96.5[d,f]	

[a]	In terms of total particulate emissions unless otherwise noted.
[b]	The scrubber purge liquor is incinerated.
[c]	The wet scrubber is a co-current system that treats 120,000 scfm of condenser off-gas. The recirculation rate of the scrubbing liquid is 5000 gpm.
[d]	Efficiency for destruction of organics.
[e]	Part of the scrubber purge liquor is treated further for maleic acid recovery.
[f]	The water scrubbers consist of a venturi contactor followed by a packed column mist eliminator.

SOLID WASTE FROM AGRICULTURE

Source	Waste	Composition
Farms, Ranches, Greenhouses	Crop waste	Cornstalks, prunings, sugarcane bagasse, drops and culls from fruit and vegetables washer waste, stubble and straw, hulls, fertilizer bags, etc.
	Forest slash	Trees, stumps, limbs, debris
Animal Husbandry	Animal manure	Lignin, organic fibers, nitrogen, phosphorus, potassium, proteins, fats, carbohydrates
	Pouch manure	
	Poultry manure	
	Swine manure	
	Animal carcasses	Ammonia and nitrates, proteins, etc., flesh, blood, fat, oil, grease
	Pesticides, insecticides, herbicides, fungicides, vermicides and microbiocides	Chlorinated organics, phosphorus, complex organic and inorganic salts

CHAPTER 4

WASTEWATER TREATMENT SYSTEMS

INTRODUCTION AND OVERVIEW

Proper handling and safe disposal of wastewater using the most economical means is sound industrial waste management. **The first step towards this objective is to characterize the wastewater for determining the best option for treatment, reuse or ultimate disposal.** Wastewater characteristics vary from industry to industry and even within the same industry due to differences in processes; technology used; age of the plant; management practices.

Industrial wastes and wastewaters are generally regarded as a problem in respect to meeting regulatory requirements. This is a result of attention given primarily to products which bring income, whereas wastes incur costs. Plant operations are generally directed at maximizing production and the resulting waste to be disposed of at minimum cost. In some cases industrial wastes might be considered a resource. Materials in the waste stream such as unrecovered products, process chemicals and cooling waters that may be recycled after treatment or recovery may lessen disposal costs.

The ability of wastewater treatment systems to perform at required levels of efficiency is critical as water pollution regulations and abatement programs have become in-

creasingly stringent. Performance, once formerly of lesser consequence, has become exceedingly important because of regulatory impacts. Improved monitoring and plant/treatment operations reduce the incidence of non-compliance. Situations where deficiencies on treatment systems occur can arise from initial design inequities or increased loads applied to the system, regardless of the cause. The result is inadequately treated discharges. Historically the solution to such a problem has been plant expansion along the same lines as the original system. Upgrading of wastewater treatment plants may be required to handle higher hydraulic or organic loads to meet existing or higher treatment requirements and loading. Existing facilities can be made to handle higher hydraulic and pollutant loads at reduced treatment efficiency by process modifications. Higher treatment requirements usually demand significant modification or expansion of existing facilities.

Upgrading an existing wastewater treatment operation may arise for one or more of the following reasons:

- Lack of proper plant operation and control

- Inadequate design

- Changes in wastewater flow characteristics

- Changes in treatment requirements

Problem areas may arise from:

- Hydraulic and organic/inorganic overloading

- Inadequate pollutant reduction/removal

- Inadequate solids removal

- Inadequate sludge handling

Wastewater treatment technology development provides innovative upgrading procedures to meet deficiencies in existing processes.

It has long been recognized that the performance of a wastewater treatment plant is affected by variations in the influent flow. **Equalization** of extreme flows can dampen the fluctuations in loading to a plant. Various processes and process equipment are marketed and successfully used to increase removal in primary and secondary clarifiers. They include the use of chemical coagulation, peripheral-feed clarifiers and inclined-tube settlers. These procedures, in many cases, have the effect of maintaining good solids removal while maximizing the hydraulic throughput in the existing facilities. Chemical addition in primary and secondary clarifiers can increase solids capture and BOD removal. In addition, several types of screening devices are available as possible substitutes for primary clarification. Modifications of the conventional activated sludge process, including step aeration, contact stabilization and complete mixing, have been adequately

studied and have been used to upgrade various treatment plants. A most significant development in the activated sludge treatment process came with the full-scale demonstration of the feasibility and effectiveness of using oxygen aeration as a substitute for air aeration.

Another method of upgrading an overloaded secondary plant is to provide additional treatment ahead of the existing biological treatment facilities. The use of plastic media trickling filters should be considered when roughing treatment is indicated. Plastic media filters have been successfully used as roughing filters in industrial wastewater treatment, and it is very likely that they will be used in the future for upgrading of treatment plants. On many occasions, treatment plants which are functioning satisfactorily (design flow not exceeded) are required to improve solids or BOD removal because of more stringent water quality standards. This additional treatment can often be achieved by polishing the treatment plant effluent. Several methods are available and have been used successfully, including aerobic and facultative lagoons, micro-straining, multi-media filtration and activated carbon treatment.

Although considerable effort has been made in the study of organic removal processes, the area of sludge handling and dewatering has also received attention. Inadequate digestion and sludge handling facilities often adversely affect overall treatment plant operation. The return of supernatant or filtrate from thickening or dewatering units to the head of the plant can impose high oxygen demands on the system and add substantial amounts of fine solids which are difficult to remove in the secondary clarifier. The high concentration of nutrients and organics in such streams and the periodic nature of the return flow often necessitate separate treatment, especially when nutrient removal is a consideration. Various sludge-handling developments which have been successful are: high-rate anaerobic digestion; aerobic digestion; thickening of sludge prior to digestion to increase the capacity of existing digesters; the use of chemicals to improve thickening and dewatering of sludges; and the use of heat treatment processes for the disposal of sludges.

WATER REDUCTION PRACTICES FOR INDUSTRIAL WASTEWATER

Volume Reduction

Classification of wastes - Separation of process wastewater, cooling water, sanitary wastewater, and rainfall run-off is necessary so that the volume of water requiring intensive treatment can be reduced.

Conservation of water - Changing from an open to a close system for process and cooling waters, good house-keeping and preventative maintenance; dry disposal of waste rather than using water for flushing, regulating water pressure, installing meters in each department to make operators cost and quantity conscious are all good practices.

Changing production to decrease waste - Improved process control, improved design, the use of different or better quality raw materials are all good practices.

Reusing industrial and municipal effluents - Adopting practices used in regions where water is scarce is an essential practice.

Strength Reduction

Processing substitutions - Use a substitute processing chemical that is less polluting (e.g., enzymes for lime and sulfides in the leather industry, cellulosic sizing agents for starch in the textile industry. H_3PO_4 for H_2SO_4 in pickling).

Equipment modifications - One example is improved designs to reduce carry over of solids to drains.

Segregation of wastes - Classification of waste by volume reduction exemplifies this practice.

By-product recovery - Recovery of chemicals, solvents and other components for recycling or reuse for other purposes, (e.g., blood from meat packing, chromium for metal plating, cellulose fibre from paper making, paint solvent from gaseous waste).

Wastewater Treatment Options and Requirements:

The segregation of waste sources includes:

- Removal of floating/settleable solids
- Removal of organic solids
- Removal of colloidal solids
- Removal of inorganic dissolved solids
- Equalization of flows and waste streams
- Neutralization of flows and waste streams

- Neutralization/pH adjustment

- Oxidation

- Chemical precipitation

- Sludge treatment

- Solid waste treatment

- Solid waste disposal

- Gaseous waste treatment

- Segregation of waste sources

TREATMENT OPTIONS FOR ORGANIC LIQUIDS

There are a wide range of physical and/or chemical processes that may be used for treating organic liquids. They may be applied to concentrated aqueous or non-aqueous mixtures and may offer materials recovery possibilities as well as applications for wastes that may be biologically difficult to treat. Such processes include:

- Adsorption
- Hydrolysis
- Air stripping
- Liquid extraction
- Centrifuging
- Oxidation-reduction
- Dialysis
- Ozonation
- Electrophoresis
- Reverse osmosis
- Evaporation
- Ultrafiltration
- Freeze crystallization

INORGANIC LIQUID TREATMENT OPTIONS

Inorganic solutions, such as metal solutions, can be treated by a number of operations which include:

- Electrodialysis
- Ion exchange
- Electrolytic recovery
- Chemical precipitation
- Evaporation
- Reverse osmosis

All of these are well known industrial processes. Here again material recovery may be possible, though expensive and not cost effective or economical.

BIOLOGICAL TREATMENT

A biological treatment process generally refers to any process utilizing micro-organisms to consume the waste. Wastes are either synthesized into new cell matter or gases. In either case dissolved substrates are removed from solution and destroyed or converted to colloidal matter. Biological wastewater treatment is most generally applied to relatively dilute organic aqueous solutions. An area under development in recent years has been an attempt to develop biotechnology of "super" microorganisms specifically designed to metabolize hazardous wastes.

ANALYSIS OF WASTE CONSTITUENTS

Chemical analyses of industrial wastewaters are performed for one or more of the following purposes:

- Estimation of material balances for processes to permit evaluation of efficiencies and to relate material losses to production operations.

- Evaluation of conformance to limits set for performance efficiency of certain processes.

- Evaluation of the effectiveness of in-plant processes, modifications, and other measures taken for reduction of losses.

- Determination of sources and distributions of waste loads for purposes of by-product recovery, segregation of flows, in relation to strength and type for separate treatment or disposal.

- Recognition of malfunctions, accidents, spills or other process disturbances.

- Determination of the type and degree of treatment required for the recovery of certain substances from waste effluents.

- Evaluation of performance for effluent quality.

- Control of treatment and discharge of waste effluents according to standards and/or according to variations in the conditions of the receiving waters.

- Provision of records of costs and compliance associated with discharge of waste effluents to municipal sewers.

PARAMETERS FOR ANALYSIS

After defining the analysis, the next step in the measurement system is to decide on constituents for which analyses are to be made and what methods are to be used.

Evaluation of wastewater characteristics might include:

- Flowrates
  Present, projected, and peak flow

- Existing treatment
  Description
  Adequacy for intended project

- Existing effluent disposal facilities
  Description
  Consideration of water rights

- Composition of effluent to be applied
  Total dissolved solids
  Suspended solids
  Organic matter (BOD, COD, TOC)
  Nitrogen forms (all)
  Phosphorus
  Inorganic ions
  Heavy metals and trace elements
  Exchangeable cations
  Boron
  Bacteriological quality
  Projected changes in characteristics

LIQUID WASTE AND WASTEWATER TREATMENT OPTIONS

This section briefly outlines treatment options for aqueous liquid wastes. These wastes are pollutants that are organic in nature.

- **Physical Operations**

Screening - Removal of large solids at the start of a treatment sequence. Screening is carried out using coarse bars or racks which can be mechanically cleaned.

Sedimentation - Removal of settleable solids by reducing the flow of the wastewater in a sedimentation tank. Discrete fast settling particles are usually removed first, followed by the slower settling flocculent particles. The insertion of inclined tubes or plates in a sedimentation tank can increase the efficiency of flocculent particle removal. This is usually referred to as a Lamella settler.

Flotation - Removal of low density solids by introducing fine air bubbles to float the solids which are subsequently skimmed. Air can be introduced through small perforations or more usually by dissolving air in the wastewater under pressure and releasing the pressure in the flotation tank.

● **Biological Processes**

Activated sludge process - The process converts organics to microbial cells, which can then be separated out. The process is carried out in an aerated tank, where wastewater is mixed with microorganisms; the mixture is then separated in a sedimentation tank, where the micro-organisms contained in flocs of solids (termed activated sludge) are sedimented from the clarified wastewater and partly returned to the aeration tank.

Trickling filtration - This process is essentially the same as the activated sludge process. The difference is that the microorganisms are immobilized on the surfaces of rocks or corrugated plastic sheets. The sloughed mass of micro-organisms are separated from the clarified wastewater in a sedimentation tank without any need for the return of the sludge.

Anaerobic digestion - The biological processes described earlier can also be carried out in the absence of oxygen, with less sludge production and with the production of methane gas. The gas can be used for maintaining the optimum digestion temperature (37°C), with any excess for other heating purposes.

CHEMICAL/PHYSICAL TREATMENT

Carbon adsorption - Carbon adsorption is used to remove dissolved organic substances which are nonbiodegradable. The clarified wastewater is commonly passed through a bed of granulated activated carbon and the operation is carried out until the adsorption capacity of the bed is nearly exhausted and the breakthrough of the organics occurs. The activated carbon can usually be regenerated by heating to a high temperature (above 600°C).

Ion exchange - Ion-exchange is used to remove cations or anions by exchanging them with cations or anions respectively attached to ion-exchange resin beads. The operation is usually carried out using beds of ion-exchanger, similar to the operation of granulated activated carbon. Since ion-exchange is reversible, an ion-exchanger can be regenerated using a brine solution, acid or alkali. The regenerated waste may require treatment prior to disposal.

Reverse Osmosis - Reverse osmosis removes dissolved substances by forcing water through a semi-permeable membrane (e.g. cellulose acetate) that does not allow the dissolved substances to pass through. The semi-permeable membrane is generally supported on a tubular frame with pressure applied to the wastewater. A concentrated stream containing over 90% of the dissolved substances is produced, which may require treatment prior to disposal.

Electrodialysis - Electrodialysis removes cations and anions by attracting them from the wastewater using an anode and cathode respectively. The operation is carried out in a cell separated alternately by cation and anion resin membranes; the ions migrating towards the electrodes form a concentrated solution in alternate compartments, which may require treatment prior to disposal.

Chemical precipitation - Many pollutants (e.g., heavy metals) can be precipitated out of solution by the addition if a suitable reagent (e.g. lime). The precipitate formed can be removed by methods used for the removal of colloidal inorganic solids.

Chlorination - Chlorination is used to remove pathogenic pollutants (bacteria and viruses), but it also removes dissolved pollutants such as iron and manganese by oxidation and precipitation.

For Aqueous Liquid Waste (Pollutants Primarily Inorganic in Nature)

- **Physical Operations:**
  Filtration, Flocculation

Sedimentation - This involves the removal of coagulated or flocculated particles by mechanical straining. The wastewater is passed through a filter medium under pressure (gravity, pumping or vacuum). The filter medium can be a sand bed or fabric filter. Colloidal particles are flocculated (coagulated) by the addition of suitable coagulating agents (e.g., lime, alum) and/or polyelectrolytes (long chain polymers with active groups which can join to the colloidal particles).

Processes that may accompany or follow include:

- Chemical process

- Sludge dewatering and treatment prior to disposal as solid waste

- Physical operations

Centrifugation - Sludge dewatering is accomplished by applying centrifugal force on the sludge thereby throwing the solids onto the wall of the centrifuge. Several types of centrifuge are available with different ways of discharging the collected solids.

- **Filtration, Thickening and Biological Processes**

Aeration - This process is used to satisfy the oxygen demand of the organics and is similar to the Activated Sludge process, but normally without sludge recycling.

Anaerobic Digestion - This involves biological treatment in an oxygen deficient solution.

- **Chemical Processes:**
  Wet air oxidation - Solid Waste

Disposal options for solid waste include incineration.

Concentrating Liquid Waste may include:

- **Physical Operation**

Evaporation - Removal of water by applying heat (e.g. solar radiation) to produce solids, which may require further treatment as solid waste.

- **Chemical Processes and Chemical Precipitation**

Oxidation/Reduction - Specific pollutants can be converted to non-hazardous forms by oxidation (e.g. cyanide) or by chemical reduction (e.g., chromate) and followed by chemical precipitation.

The solid waste disposal options include incineration and solidification.

ACTIVATED SLUDGE PROCESSES

Basically, in the activated sludge process microorganisms in suspension are used to oxidize soluble and colloidal organics to CO_2 and H_2O in the presence of molecular oxygen. During the oxidation process, a portion of the organic material is synthesized into new cells. A part of the synthesized cells then undergoes auto-oxidation in the aeration tanks, the remainder forming excess sludge. Oxygen is required to support the oxidation and synthesis reactions. In order to operate the process on a continuous basis, the solids generated must be separated in a clarifier for recycling to the aeration tank, with the excess sludge from the clarifiers being withdrawn for further handling and disposal.

Conventional Activated Sludge - The wastewater is commonly aerated for a period of 6 to 8 hours (based on the average design flow) in the presence of a portion of the secondary sludge. The rate of sludge return expressed as a percentage of the average wastewater design flow is normally about 25 percent, with minimum and maximum rates of 15 and 75 percent. The plug flow mixing configuration is used to condition the biological organisms for improved clarification. This is accomplished in rectangular tanks, designed so that the total tank length is generally 5 to 50 times the width.

The following factors have been cited as limitations in the design and use of the conventional activated sludge process:

- BOD loadings are limited to about 35 lbs./1,000 cu. ft./day

- A high initial oxygen demand is experienced in the head end of the aeration tank.

- The final clarifier is subjected to high solids loadings.

- It is necessary to increase sludge recirculation proportionately with increasing BOD loadings.

- Detention times are in the range of 6 to 8 hours.

- There is a lack of operational stability with variations in hydraulic and organic loadings.

Some of these limitations have stimulated the development and use of various activated sludge modifications, such as step aeration, contact stabilization, completely-mixed, two-stage activated sludge and the use of oxygen aeration instead of air as a source of dissolved oxygen.

TRICKLING FILTER PROCESSES

Trickling filtration consists of a uniform distribution of wastewater over the trickling filter media by a flow distributor. A large portion of the wastewater applied to the filter rapidly passes through it and the remainder slowly trickles over the surface of the slime. BOD removal occurs by biosorption and coagulation from the rapidly moving portion of the flow and by progressive removal of soluble constituents from the more slowly moving portion of the flow. The quantity of biological slime produced is controlled by the available food and the growth will increase as the organic load increases until a maximum effective thickness is reached. This maximum growth is controlled by physical factors including hydraulic dosage, rate, type of media, type of organic matter, amount of essential nutrients present and the nature of the particular biological growth.

High-rate trickling filters have hydraulic loadings of 10 to 30 mgpd and organic loadings up to 90 lbs. BOD/1,000 cu.ft./day including recirculation. In all high-rate filters, some form of recirculation is used in order to maintain a relatively constant hydraulic loading. The correspondingly higher loadings result in an overall BOD removal efficiency that is somewhat lower than that obtainable from a low-rate trickling filter. The higher organic loadings in high-rate filters preclude the development of nitrifying bacteria in the lower section of the filter. Hence, these plants will seldom exhibit any incipient nitrification.

Super-rate trickling filters have evolved as a result of the development of various types of synthetic trickling filter media. Past experience has indicated that hydraulic loadings of 150 mgpd and higher, including recirculation, may be accommodated in super-rate trickling filters. There are numerous factors that affect the performance of trickling filters. Some of these are:

- Wastewater Characteristics
- Trickling Filter Media
- Trickling Filter Depth
- Recirculation
- Hydraulic and Organic Loading
- Ventilation
- Temperature of Applied Wastewater

EFFLUENT POLISHING TECHNIQUES

The use of effluent polishing for secondary effluent has received attention as a practical and economical method of upgrading to obtain increased organic and suspended solids removal from existing treatment facilities. It appears to be particularly applicable in those cases where it is necessary to increase efficiency by an overall amount of 10 to 20 percent in order to meet stricter water quality standards. Four unit processes are considered for effluent polishing: polishing lagoons; microstraining; filtration, including mixed, multi-media and moving-bed filters; and activated carbon adsorption.

Polishing lagoons offer an opportunity for increased organic and solids removal at a minimum cost. There are two types of polishing lagoons which can be used, aerobic and facultative. Aerobic lagoons are generally subdivided into two groups:

- Shallow lagoons, with depths in the range of 2.5 to 4.0 feet.

- Deep lagoons, with aeration devices included to insure maintenance of aerobic conditions.

ACTIVATED CARBON ADSORPTION

The limitations of conventional biological treatment processes in regard to reliable achievement of a high degree of organic removal (particularly of certain compounds which are refractory to biodegradation), along with increasingly strict water quality standards, emphasize the need for a supplementary organic removal process. Activated carbon has been used to provide tertiary treatment of biologically treated effluents. Moreover, experience gained from the operation of activated carbon plants for tertiary treatment of wastewater suggests that activated carbon need not be restricted to a polishing role, but can be used as an alternative to biological treatment. Activated carbon for wastewater treatment can be used either in the powdered or in the granular forms. The impracticality of economical regeneration has restricted the use of powdered carbon in wastewater treatment.

The adsorption of organic materials from wastewater onto the activated carbon involves complex physical and chemical interactions. Biological degradation of adsorbed materials also occurs and this can significantly enhance the overall treatment performance. The ability of activated carbon to adsorb large quantities of dissolved materials from wastewater is due to its highly porous structure and to the resulting large surface area, which provides many sites for adsorption of dissolved materials.

Important factors in the design of activated carbon treatment facilities include: pretreatment requirements; particle size; hydraulic loading and contact time; regeneration losses; flow configuration; and required effluent quality.

Treatment of wastewater by activated carbon requires that the influent total suspended solids concentration be less than about 50 mg/l. This is essential in order to use the activated carbon bed as an adsorption medium and to minimize its filtration function. If the solids loading is much higher than 50 mg/l, a filter may be needed in advance of downflow carbon beds, or upflow carbon beds may be required for operation.

Theoretically, carbon particle size primarily affects the rate of adsorption and not the capacity of the carbon. Adsorption rates are greater for smaller particle sizes than for larger particle sizes. However, adsorbents close to saturation will be less affected by particle size than adsorbents in their virgin state. Contact time, hydraulic loadings and

bed depth are interrelated physical parameters. Of the three, contact time is clearly the most important. Since the activated carbon treatment of wastewater requires that a definite contact time be established to complete the adsorption process, any increase in applied hydraulic load necessitates a deeper carbon column to maintain the same contact time.

DISINFECTION AND ODOR CONTROL

Disinfection and odor control are two areas which have received attention from regulatory agencies through the establishment and enforcement of rigid bacteriological effluent standards and air pollution standards.

Wastewater plants in the United States have used chlorination for disinfection purposes. To be effective for disinfection purposes, a chlorine residual of 0.2 to 1.0 mg/l is recommended with a contact time of not less than 15 minutes at peak flow rates.

Odors from wastewater treatment plants can usually be attributed to three sources: septic raw wastewater, overloaded secondary treatment facilities and sludge treatment practices. Septicity in wastewaters is caused by the depletion of dissolved oxygen due to long residence in sewers and the subsequent increase in anaerobic activity. As wastewater becomes anaerobic, facultative and anaerobic bacteria flourish. These bacteria utilize nitrates and sulfates present in wastewater as their oxygen source. The reduction of sulfate ions produces the highly odorous gas, hydrogen sulfide. Other odorous gases which may be present are indole, skatole, mercaptans, disulfides, volatile fatty acids and ammonia. Increased summer temperature and extended detention times can result in the rapid build-up of hydrogen sulfide and carbon dioxide. At a pH level below 8, the equilibrium shifts toward the formation of non-ionized H_2S and is about 80 percent complete at pH 7. At pH 8 and above, most of the reduced sulfur exists in solution as HS^- and $S^=$ ions. H_2S is noticeable even in the cold when present in water to the extent of 0.5 mg/1. When present to the extent of 1.0 mg/1, it becomes very offensive.

Overloaded secondary treatment facilities are also a potential source of odor. If the air supply to an activated sludge aeration tank is inadequate, odorous conditions usually develop. It is also possible that a properly sized air supply system can strip odorous gases from septic wastewater.

Odors associated with sludge treatment occur in thickening, digestion and sludge dewatering facilities. Thickeners may receive both septic primary and secondary sludges. Gases from well-operated digesters may contain small quantities of H_2S which are usually destroyed by normal flaring of digester gas. The predominant odor in digested sludge is ammonia, although traces of volatile organic acids may be present.

Various methods available for control of odors emanating from a wastewater treatment plant are:

- Changes in the operational procedures and new techniques.

- Chemical treatment or pre-treatment, which might include chlorine, ozone, lime or powdered carbon.

- Collection and treatment of noxious gases.

Chlorination is used for two purposes: to retard biological action which produces odors and to react chemically with odorous sulfur compounds, oxidizing them to innocuous sulfur forms, usually free colloidal sulfur.

Ozone has been added to wastewaters for odor control with some favorable results. Because of the extremely high reactivity of ozone, a much higher ozone demand is generally exhibited by a wastewater than would be exhibited for chlorine. However, the use of oxygen aeration in secondary treatment may have an added benefit, since the exhaust gas could provide the ozone generator with an economic source of oxygen. Due to the high cost of ozone generation, the use of ozone for odor control has been limited.

Lime and powdered carbon have also been used in various applications for odor control. The addition of lime to septic wastewater raises the pH. Since the solubility of H_2S increases with increasing pH, less H_2S evolves, thereby decreasing the odor level. Powdered activated carbon adsorbs odor-causing materials and, thereby, decreases the odor level.

SLUDGE THICKENING

Use of **air flotation** for upgrading is limited primarily to thickening of sludges prior to dewatering. Used in this way, the efficiency and/or capacity of the subsequent dewatering units can be increased and the volume of supernatant from the following digestion units can be decreased. Existing air flotation thickening units can be upgraded by the optimization of process variables and by the utilization of polyelectrolytes. **Air flotation** thickening is best applied to thickening waste activated sludge. With this process, it is possible to thicken the sludge to 6 percent, while the maximum concentration attainable by gravity thickening without chemical addition is 2-3 percent. The air flotation process can also be applied to mixtures of primary and waste activated sludge. The greater the ratio of primary sludge to waste activated sludge, the higher the permissible solids loading to the flotation unit. Due to the high operating costs, it is generally recommended that air flotation be considered only for thickening waste activated sludge.

The most commonly used type of air flotation unit is the dissolved air pressure flotation unit. In this unit, the recycled flow is pressured from 40 to 70 psig and then saturated with air in the pressure tank. The pressurized effluent is then mixed with the influent sludge and subsequently released into the flotation tank. The excess dissolved air then separates from solution, which is now under atmospheric pressure and the minute (average diameter 80 microns) rising gas bubbles attach themselves to particles which form the sludge blanket. The thickened blanket is skimmed off and pumped to the downstream sludge handling facilities while the subnatant is returned to the plant.

Gravity thickening is the most common process in use today for the concentration of sludge prior to digestion and/or dewatering. Thickeners can contribute to the upgrading of sludge handling facilities as follows:

- Increase the capacity of overloaded digesters or subsequent sludge handling units.
- Reduce the size and increase the efficiency of sludge digestion or dewatering units.
- Improve primary clarifier performance by providing continuous

withdrawal of sludge, thereby insuring maximum removal of solids.

The process is simple and is the least expensive of the available thickening processes. The reduction in size and improvement in efficiency of subsequent sludge handling processes often can offset the cost of gravity thickening. The process also allows equalization and blending of sludges, thereby improving the uniformity of feed solids to the following processes. Existing gravity thickeners can be upgraded by providing continuous feed and drawoff, by diluting the feed solids and by chemical addition.

Gravity thickening is characterized by zone settling. The four basic settling zones in a thickener are:

- The clarification zone at the top containing the relatively clear supernatant.
- The hindered settling zone where the suspension moves downward at a constant rate and a layer of settled solids begins building from the bottom of the zone.
- The transition zone characterized by a decreasing solids settling rate.
- The compression zone where consolidation of sludge results solely from liquid being forced upward around the solids.

SLUDGE DEWATERING

Vacuum Filtration - Process operating considerations for vacuum filtration include:

- Control of feed solids concentration.

- Chemical requirements for sludge conditioning.

- Sludge mixing and flocculation.

- Drum speed and drum submergence.

- Filter fabric characteristics.

Each of these parameters affects the filter yield, economy of operation and filter cake characteristics. The feed solids concentration can be controlled by pre-thickening the şludge. In general, the higher the feed solids concentration, the higher the filtration rate and the filter yield. Few, if any, raw or digested wastewater sludges can be successfully dewatered without some form of chemical conditioning using ferric chloride, lime and/or polyelectrolytes. Proper sludge conditioning requires a determination of optimum chemical dosages. Experience and careful laboratory monitoring of the sludge characteristics are key factors in maintaining the proper chemical proportions and concentrations.

Centrifugation - Centrifuges have been used for many years by various industries for clarifying liquids, concentrating solids, separating immiscible liquids and purifying oils. However, their use in the wastewater field for sludge dewatering is not as widespread as is the use of vacuum filters. Improvements in centrifuge design, efforts by the centrifuge industry to enter the wastewater treatment field and broader dissemination of centrifuge performance data have encouraged increased use of centrifuges for thickening and dewatering of primary, secondary and combined wastewater sludges. Centrifuges have good potential for upgrading overloaded solids handling facilities due to their flexibility in operation and lesser space requirements compared to vacuum filters. There are three general classifications of centrifuges that can be applied to sludge thickening and dewatering: solid-bowl, disc and basket centrifuges. The most popular type of centrifuge today is the solid-bowl because of its dependable performance and low maintenance requirements. The solid-bowl machine has a spinning cylinder which causes particles to settle out along its inner wall. Solid-bowl centrifuges are especially suited to dewatering primary wastewater sludge and mixtures of primary and waste biological sludge. They are also able to dewater waste biological sludge alone, but some form of polymer addition is required in order to operate at an economical feed rate and to obtain solids concentrations above 5 or 6 percent. For most sludges, to achieve solids recovery in the range of 80 to 95 percent with a solid-bowl centrifuge requires the addition of polymers to the sludge.

The basket centrifuge is a tubular type centrifuge with a solid bowl and, therefore, is similar to the solid-bowl centrifuge in that the solids settle out along the inner wall due to centrifugal force. The solids are removed on an automated batch basis. Because of the large bowl diameter, the basket centrifuge is operated at slower speeds. The centrifuge can be operated on automatic cycle for programmed filling and skimming. Abrasion occurs only with the skimming tool. Hence, for the most part, this is a low-speed, low-maintenance unit. The application of basket centrifuges to the wastewater treatment field is relatively new.

The reader should review the process flowsheets at the end of this chapter to become familiar with various types of separation equipment and processes.

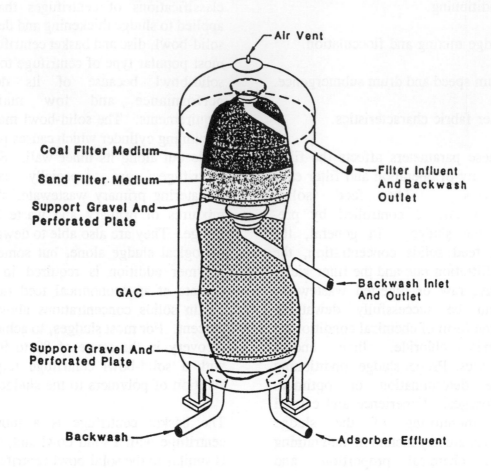

Air Vent

Coal Filter Medium

Sand Filter Medium

Support Gravel And
Perforated Plate

GAC

Support Gravel And
Perforated Plate

Backwash

Filter Influent
And Backwash
Outlet

Backwash Inlet
And Outlet

Adsorber Effluent

DUAL-STAGE FILTER/ADSORBER SYSTEM

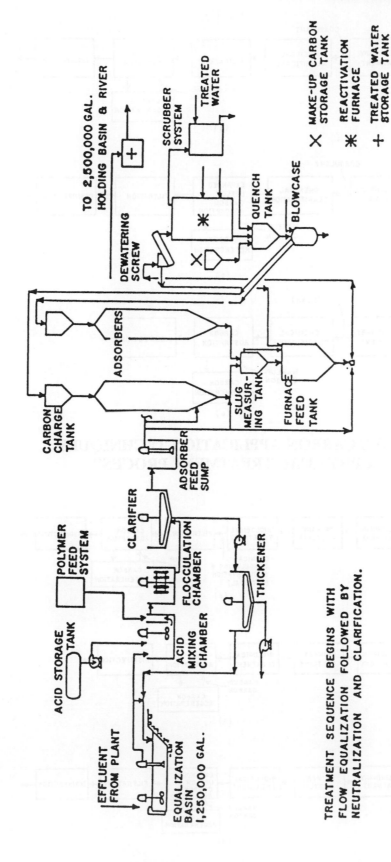

FLOW SCHEMATIC OF WASTEWATER TREATMENT FACILITY

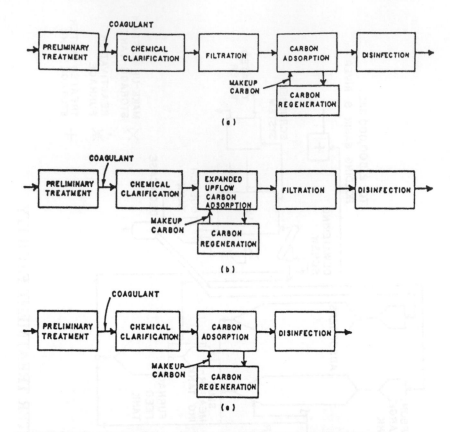

VARIOUS CARBON APPLICATION TECHNIQUES IN A PHYSICAL TREATMENT PROCESS

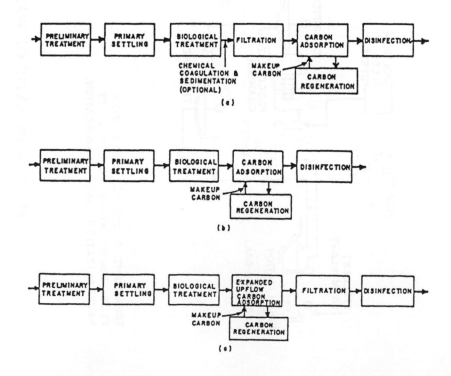

POINTS OF APPLICATION OF CARBON IN A BIOLOGICAL TREATMENT PLANT

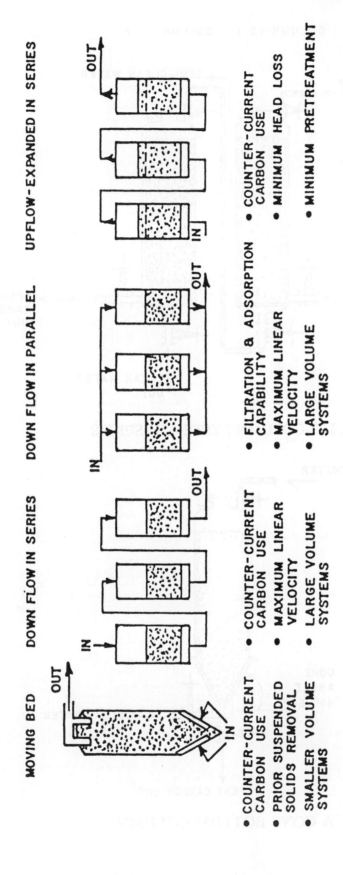

TYPICAL CARBON ADSORPTION SYSTEMS AND THEIR ATTRIBUTES

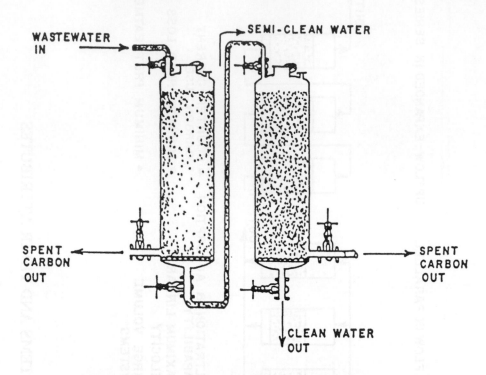

COLUMN NO. 1   COLUMN NO. 2

WASTEWATER IN →

SEMI-CLEAN WATER →

SPENT CARBON OUT ←

→ SPENT CARBON OUT

CLEAN WATER OUT

MULTI-COLUMN SYSTEM IN SERIES

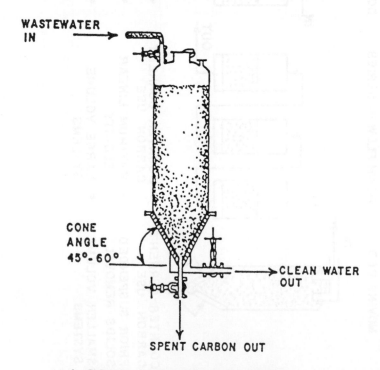

WASTEWATER IN →

CONE ANGLE 45°-60°

→ CLEAN WATER OUT

SPENT CARBON OUT

A CONE BOTTOM COLUMN

CHAPTER 5

HAZARDOUS MATERIALS AND
SLUDGE TREATMENT, DISPOSAL METHODS

INTRODUCTION AND OVERVIEW

In continuing to meet mandated guidelines by Congress, EPA has restricted land disposal of nearly one third of all hazardous wastes. Regulation as required by the 1984 RCRA amendments restricts disposal of an estimated 861 million gallons of a wide range of industrial wastes and is estimated to cost $950 million a year for companies to comply. The land ban rule which hikes chemical waste disposal costs was published in the Federal Register and requires prior treatment for specific standards for each and recommends specific technologies that can be used to meet standards.

Treatment standards for waste streams vary depending on pollutants. For most wastes of high organic chemicals concentrations, the standards are based on incineration. If metals are the problem, as in the metallurgical industries, standards are based on metals recovery or stabilization.

Electroplating sludges, another major waste, must be stabilized, which can be accomplished by using cement kiln dust, EPA says.

EPA has not specified treatment standards for wastes commonly produced by some manufacturing processes and by certain industries, or for some very small volume discard commercial chemical products and spill residues allowed under RCRA. However, EPA stipulates if treatment is available, such wastes have to be treated prior to landfill disposal.

Originally, it was thought that the petroleum refining waste streams market is an estimated $17-billion and even more impressive is the fact that these businesses are expected to increase. In the area of environmental/hazardous waste control, several processes have been developed that do more than lower disposal/compliance costs for wastes and by products: They actually transform wastes into salable by-products. Traditionally in considering separations of solids from liquids technologies such as filtration/crystallization/ extraction/evaporation/distillation, or other such familiar procedures were reviewed. Times have changed, for one thing, some mixtures cannot be separated by conventional methods and require new technology. Moreover, the goal for the most economical technology leads to careful consideration to evaluate several options before deciding which one to use. Capital and operating costs are carefully examined and ultimately selection is based on the best combination of cost effectiveness, performance and regulatory compliance.

WHAT IS SLUDGE?

The increased emphasis placed on water pollution and hazardous waste control caused by stringent regulations has resulted in the increase in solids to be treated and disposed, particularly from wastewater. Wastewater solids include grit, scum, oily material and biological sludge.

Suspended solids, oil, scum and grit are generally removed by gravity settling, skimming or flotation. These solids form the basis of what is known as primary sludge because they are removed in the primary section of the wastewater treatment plant and, in general, the settleability of these solids makes them easier to treat.

Biological sludge, also known as secondary sludge, is generated by biological treatment processes, employing microorganisms to remove BOD and COD from wastewater. Secondary sludge is more difficult to treat because of inherently poor settling qualities. Additionally, chemical sludges, formed by the addition of chemicals and subsequent settling, are generated in primary/secondary/tertiary sections of wastewater treatment plants.

Sludge removal and subsequent treatment did not receive a great deal of attention in the early days of wastewater treatment. The advent of stricter solid and hazardous wastes disposal regulations has required selection of sludge treatment methods and ultimate disposal is increasingly important. Many options for sludge treatment and disposal are available.

SLUDGE PRODUCTION AND CHARACTERIZATION

Primary Sludge

Primary sludge is formed when easily settleable solids are removed by primary sedimentation processes. Primary sludge normally makes up about half of the sludge produced in a wastewater treatment plant, while sludge formed in biological and chemical processes composes the other half of the sludge. Primary sludge is usually easier to handle than chemical and biological sludges since it is readily thickened by gravity, it can be mechanically dewatered, it forms a drier cake, and it facilitates better solids capture when a dewatering device is employed.

Primary sludge production in typical wastewater treatment plants is in the range of 800 to 2,500 pounds per million gallons of wastewater. Sludge production is best estimated by determining the total suspended solids (TSS) loading to the primary sedimentation unit and assuming a removal efficiency for that unit. When accurate TSS loading data is unavailable, an estimate of 0.15 to 0.24 pounds per capita per day is commonly used in the design of municipal wastewater treatment plants. Solids settling efficiency for primary sedimentation tanks is in the range of 50 to 65 percent, usually 60 percent for municipal wastewater sludge containing no industrial contaminants, no chemical coagulants or flocculants, no secondary or tertiary treatment sludge nor any major sidestreams from sludge processing. Combining of industrial wastewater with municipal wastewater generally increases the settleability of the suspended solids in the primary sedimentation unit. Increased settleability leads to increased sludge production.

Primary sludge characteristics and composition varies with the characteristics of the influent wastewater. The volatile solids content of the primary sludge is approximately the same as the volatile solids content of the wastewater. Primary sludge also contains grit which has escaped grit removal chambers, anaerobic and facultative species of bacteria, organics and nutrients such as nitrogen and phosphorus.

BIOLOGICAL SLUDGE

Biological sludge is produced by secondary treatment operations such as trickling filters, activated sludge processes, and rotating biological contactors. The composition and characteristics of biological sludges varies widely and is a function of the growth rate and metabolism of microorganisms in the sludge. The quality of biological sludge as well as the quantity produced is also dependent upon the treatment operations which precede the secondary treatment system. When no primary clarification is available, the biological sludge usually contains a large amount of solids such as grit, which are normally found in the primary sludge. If primary clarification is present, the biological sludge will not contain many of the components normally found in the primary sludge. Biological sludge is more difficult to dewater or thicken than both primary and chemical sludges.

There are many models available for the prediction of sludge production from activated sludge processes. Sludge production in this process is dependent on variables such as the feed composition of the waste, the amount of dissolved oxygen in the wastewater and the temperature of the wastewater. In general, activated sludge contains higher amounts of nitrogen, phosphorus and protein than primary sludge.

On the other hand, the amount of grease, fats, cellulose and the specific gravity of activated sludge is lower than that of primary sludge.

Trickling filters as used primarily in municipal waste treatment applications, forms sludge that is very similar in composition to that formed in the activated sludge system, however, different models for the prediction of the amount of sludge generated have been developed. The total mass of microorganisms present in a trickling filter system is proportional to the media surface area available, thus volatile solids production in the sludge can be calculated. Microorganisms inhabiting a trickling filter are complex and include many species of algae, bacteria, fungi, protozoa, worms, snails and insects.

Rotating biological contactors are used to remove BOD and suspended solids as are trickling filters and the activated sludge process. In this process, wastewater comes into contact with microorganisms present on plastic media in the form of discs which rotate alternately through the water and the air. Excess bacteria sloughs off the media and into the water which subsequently flows to a sedimentation tank where the sludge is settled. The composition and characteristics of rotating biological contactor sludge is not as well documented as that of trickling filter and activated sludge process sludges. However, rotating biological contactor sludge product has been estimated at 0.4 to 0.5 pounds of total suspended solids per pound of BOD_5.

CHEMICAL SLUDGE

Chemical sludges are formed when chemicals are added to wastewater for purposes such as coagulation and resulting suspended solids removal as well as for phosphorus removal. Some wastewater treatment plants apply chemicals in their tertiary treatment section to facilitate the removal of chemical precipitates, while other chemical sludges are formed in the primary or secondary treatment section to facilitate the removal of chemical precipitates, while other chemical sludges are formed in the primary or secondary treatment sections. More commonly, chemicals are added to sedimentation or biological processes, therefore, the chemical sludge formed is mixed with primary or biological sludge. Chemical sludges formed from lime addition to wastewater in the primary treatment section are easier to dewater than those formed in combination with waste activated sludge. Lime addition usually results in a sludge that thickens and dewaters better than the same sludge without chemicals. Sludges which contain metals such as iron and aluminum do not dewater as well as nonchemical sludges. However, iron containing sludges dewater more easily than do those containing aluminum. On the other hand, addition of aluminum to activated sludge makes it easier to thicken than activated sludge without chemical addition. Thickening and dewatering properties of chemical sludges can be further improved by the addition of anionic polymers.

TRACE ELEMENTS IN SLUDGES

Both organic and inorganic trace elements can be found in wastewater sludges. Trace elements entering a treatment plant must also leave that plant in either a gaseous, liquid or solid form. Trace elements are present in both industrial and municipal wastewaters. In addition to those trace elements entering the plant in the wastewater, trace elements are derived from chemicals used in wastewater treatment and

sludge conditioning processes. Ferric chloride is an example of a chemical used in the treatment of wastewater solids.

Removal of trace elements from wastewater sludges is not always practical, therefore, these trace elements must be controlled at their source. A major source of trace elements must be controlled at their source. A major source of trace elements is the metal finishing industry which produces such by-products as cadmium, chromium, copper, lead, nickel and zinc. Control of these metals at their source will significantly reduce the concentrations of these trace elements in the sludge from metal finishing wastewater treatment plants. Toxicity of trace elements to living organisms makes them a significant environmental problem. Trace elements can be reduced to a less toxic form in some wastewater treatment unit operations. An example of such an operation is when highly toxic hexavalent chromium is reduced to a less toxic form, namely trivalent chromium, in secondary treatment processes.

WASTEWATER SOLIDS

In addition to those sludges generated in wastewater treatment processes, other solids are removed from the influent wastewater before its final discharge. These solids must be removed from the wastewater in order for it to meet the local effluent standards. Among these other solids found in wastewater are screenings, grit, scum, septage and backwash. Such miscellaneous solids often find their way into relatively pure primary, secondary or chemical sludges. Mixing of miscellaneous solids with these sludges is more difficult to handle and treat. Furthermore, their addition will make recovery of valuable compounds from the sludges uneconomical, if not impossible. As a result, it is beneficial for treatment plant operators to segregate these solids from the primary, secondary and chemical sludges generated by the plant.

SLUDGE TREATMENT

Thickening

Sludge thickening refers to a class of unit operations designed to increase the solids concentration of a dilute sludge from its initial value to some higher value; usually to about 10 to 12 percent total solids. Among the unit operations employed in thickening of sludge are sedimentation, flotation and centrifugation, as well as other operations which are less commonly used. While thickening of sludge produces a material which is still a fluid, its main purpose is to achieve volume and weight. A typical gravity thickener is illustrated in **Figure 1**.

The choice of a unit operation for sludge thickening is dependent on the characteristics of the sludge to be treated, sludge processing following thickening and the type and size of the wastewater treatment facility. In general, gravity thickening processes such as sedimentation, are used for primary sludges or primary sludges combined with secondary sludges, while flotation and centrifugation are normally used on secondary sludges.

Sedimentation

Sedimentation is commonly used to thicken primary sludges. The primary clarifier itself can be used as a thickener under certain circumstances. Primary sludge thickens very well when certain conditions are met, namely the sludge is fresh, the sludge contains low amounts of biological solids and

the wastewater is reasonably cool. Thickening of secondary sludges by sedimentation is much more difficult because biological sludges are difficult to thicken by gravity, therefore this process is not as widely used in secondary sludge applications.

Sedimentation follows the principals of gravity settling. The general term settling describes all types of particle fall, solid, liquid and gas through a fluid medium. On the other hand, the term sedimentation is used to describe settling phenomena in which the particles or aggregates are suspended by hydrodynamic forces only, compression being absent. Thus, sedimentation denotes a process similar to clarification in which particles are settled by gravity in order to increase the solids concentration of a sludge. Sedimentation works best when used for sludges which are easily settleable such as primary sludge. **Figure 2** illustrates typical clarifier configurations and **Figure 3** shows rectangular sedimentation tanks.

Flotation

A second method used for sludge thickening is flotation. Flotation is also a process used to separate solid particles from a liquid phase, however, instead of settling to the bottom of the solution, the solids are carried to the surface. Flotation of solids is facilitated by the introduction of fine air bubbles into the system. Particles with a density greater than water as well as with a density less than water can be removed by flotation. Dissolved air flotation, a process in which fine air particles (50 - 100Å) which are under pressure are introduced into the wastewater to carry solids to the surface, is commonly used for thickening of hard to settle biological sludges. Although this process is more suitable for the thickening of secondary sludges, models have also been

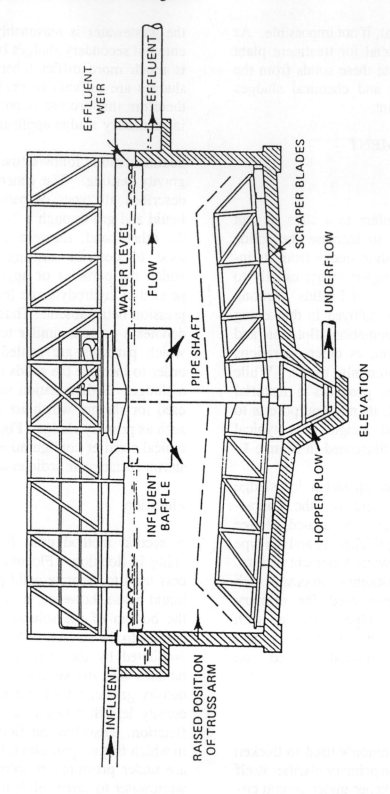

Figure 1.   Illustrates a gravity thickener

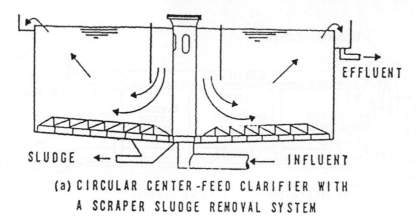

(a) CIRCULAR CENTER-FEED CLARIFIER WITH
A SCRAPER SLUDGE REMOVAL SYSTEM

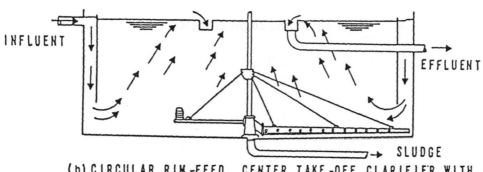

(b) CIRCULAR RIM-FEED, CENTER TAKE-OFF CLARIFIER WITH A
HYDRAULIC SUCTION SLUDGE REMOVAL SYSTEM

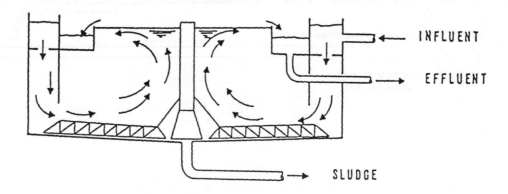

(c) CIRCULAR RIM-FEED, RIM TAKE-OFF CLARIFIER

Figure 2.   Typical clarifier configurations

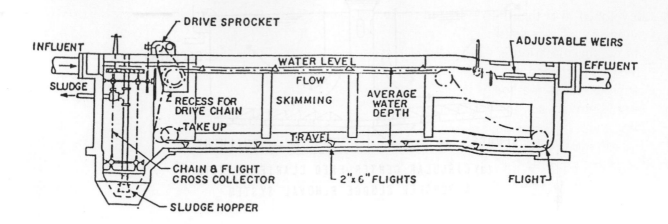

A.   With chain and flight collector

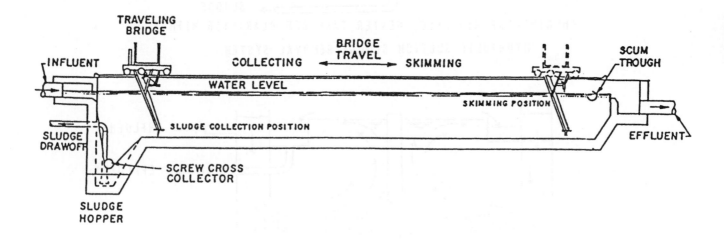

B.   With traveling bridge collector

Figure 3.   Illustrates rectangular sedimentation tanks

developed for the thickening of activated sludge by gravity settling. Test results have shown that flotation is an efficient means of thickening secondary, hard to settle sludges. However, this method is more costly than simple gravity settling since the air must be kept under pressure until it is released in the flotation vessel. **Figure 4** illustrates a dissolved air flotation system.

Centrifugation

Sludge thickening using a centrifuge was not commonly considered until after 1960. Improvements in centrifuge maintenance procedures have made it a more viable option, although it is still not as popular as gravity settling or flotation. Centrifugation is an acceleration of the sedimentation process through the use of centrifugal force. The rotating bowl of the centrifuge acts as a highly effective settling tank. Solids are collected in the bowl while the water is removed and recycled back to the water treatment unit operation. Like flotation, centrifugation is used on hard to settle secondary treatment sludges and has higher capital and operating costs associated with it than does gravity settling. Unlike flotation, centrifugation is not common practice although it is sometimes considered.

Other Methods

Other methods considered in sludge thickening applications are elutriation, secondary anaerobic digestion, facultative sludge lagoons and ultrafiltration.

Elutriation is a process used for washing and thickening digested primary sludges. This method is also effective for the thickening of a combination of primary and biological sludges when a flocculant is used to co-flocculate the mixed sludges.

Secondary anaerobic digestors serve as good holding tanks for secondary sludges, however, generally they do not function well as thickeners. Facultative sludge lagoons are not favored in thickening applications either, even though they are an effective means of further concentrating an aerobically digested sludge. Thickening of sludge of ultrafiltration has been studied but it has never been applied industrially. Results of studies show that ultrafiltration requires both a high pressure drop and high utility consumption to achieve reasonably good sludge thickening.

Stabilization

Stabilization is a process used to reduce the amount of odor produced by a sludge as well as the pathogenic organism content. This process is used mainly when sludge is to be further treated by conventional sludge treatment unit operations before ultimate disposal. When the sludge treated is to be incinerated, no stabilization is necessary since odorous compounds will be oxidized and pathogenic microorganisms destroyed at high temperatures.

Stabilization of waste sludge is accomplished by four basic processes, anaerobic digestion, aerobic digestion, lime stabilization and chlorine oxidation. Both anaerobic and aerobic digestion are increasing in popularity. Production of methane gas in the anaerobic digestion process has made it increasingly more attractive. Aerobic digestion produces a liquid effluent of high quality and potentially exothermic reaction conditions. Lime treatment, anaerobic and aerobic digestion are all considered when utilization of sludge rather than disposal is the process goal. Chlorine oxidation is used only in limited situations, mainly where septic tank sludges are involved.

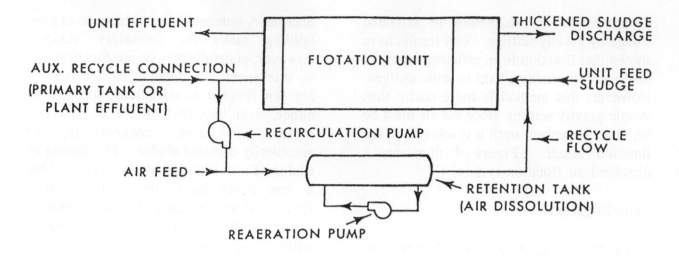

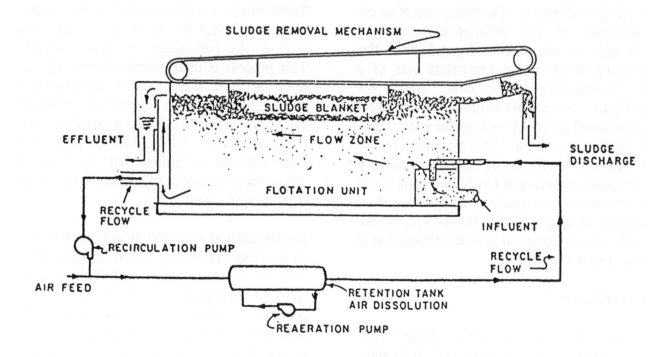

**Figure 4.   Top figure illustrates the dissolved air flotation system.
The bottom figure is a schematic of a dissolved air flotation unit.**

Anaerobic digestion is the biological degradation of complex organic substances in the absence of free oxygen. As a result of this process, organic material is converted to methane, carbon dioxide and water while energy is released. Release of energy combined with the formation of methane from carbon containing materials leaves nothing to sustain biological life, therefore all remaining solids are rendered stable.

Anaerobic digestion is an old form of stabilization used for many years in the treatment of sewage sludges. At one time, anaerobic digestion took place in the same tank that was employed for sedimentation. Further developments in the process necessitated the use of a separate tank for better stabilization. Development of digester heating, together with mixing in the reactor have generated further process improvements making it a most popular form of sludge stabilization. Aerobic digestion uses oxygen for the biochemical oxidative stabilization of wastewater sludge in an open or closed tank. Early studies were done on the aerobic digestion for the removal of volatile solids.

Advantages of the aerobic digestion process include lower capital costs than those associated with anaerobic digestion; ease of operation; no generation of nuisance odors; production of a supernatant low in BOD_5, suspended solids and ammonia nitrogen; reduction in the quantity of grease or hexane solubles in the sludge mass; and reduction of the level of pathogens in the sludge. Disadvantages include the production of a digested sludge with poor dewatering characteristics, high power costs for oxygen supply, and the influence of temperature on system performance.

Lime stabilization of sludges is a relatively simple process. Lime is added to the sludge in order to raise the pH to a level of 10 to 11. Increasing the pH to this level significantly reduces odors generated by the sludge as well as the biological activity of potentially pathogenic microorganisms. An advantage of this process over other stabilization processes is that capital and operating costs are low. On the other hand, stabilization of the sludge is rarely complete when lime is added since reduction in pH below 10 or 11 downstream leads to resumption of odor production and increased biological activity. Furthermore, lime addition leads to chemical precipitation, creating more solids to be handled and disposed of later on.

Chlorine oxidation of wastewater sludges produces a sludge which can and has been used for soil conditioning of croplands. Chlorine oxidation is a rapid reaction, therefore, equipment sizes are drastically reduced when employing this operation. On the other hand, chlorine oxidation produces a very acidic sludge, therefore acid-resistant materials of construction are required in this process. In addition, like aerobic digestion and lime stabilization, chlorine oxidation does not produce a methane by-product which can be used for energy production. As a result, anaerobic digestion remains the most popular process for sludge stabilization.

Disinfection

Disinfection of wastewater sludges is carried out with the intention of destroying or inactivating pathogenic microorganisms. Destruction is the physical disruption of pathogenic microorganisms, while inactivation is the removal of a pathogen's ability to infect. Disinfection is an important step in the sludge treatment process since potential pathogenic organisms must be removed before sludge can be used for soil treatment of croplands, forests or parks. The type of system chosen for sludge disinfection should be designed to produce a sludge suitable for its final use or destination.

Pathogens are biological agents capable of causing disease in host organisms. The host organism can be human, animal or even a plant. Pathogens can be classified in one of four broad categories; viruses, bacteria, parasites and fungi. Viruses, bacteria and parasites are present in all municipal wastewater sludges as a result of human feciation upstream of the wastewater treatment plant. Fungi, on the other hand, grow in wastewater sludges which have been stored or allowed to sit for extended periods of time. The number of pathogens in a wastewater sludge is significantly reduced during the stabilization process. However, some organisms, particularly parasites are able to survive the stabilization process and therefore, necessitate the disinfection of the wastewater sludge. Data on the survival of parasites during the stabilization process is sparse. Some studies indicate that parasites such as protozoa cysts should not survive the stabilization processes, while other protozoa like the helminth ova survive these processes.

Chemical treatment of wastewater sludges is used as a disinfection technique. Among the chemicals used for disinfection are lime and chlorine. Lime disinfection is effective if the pH of the sludge is maintained above a level of 10 to 11. Studies done on the use of lime for stabilization indicated increased pathogenic bacteria reductions when the pH of the sludge reaches values of 11.5 and 12.5. Chlorine is an effective disinfectant when applied in large amounts to leave a re-

sidual concentration. Chlorine readily inactivates both bacteria and viruses, however, it is relatively ineffective against cysts and ova of parasites.

Other methods which can be used for the disinfection of wastewater sludges include heat treatment and high energy radiation. Heat treatment processes have long been used for disinfection. One such process, pasteurization, is used for the disinfection of foods suitable for human consumption. Some pathogens are resistant to heat and will survive the pasteurization process. For instance, studies conducted have proven that fecal streptococci are the most heat resistant among bacteria, followed by coliforms and salmonella. Use of high energy radiation in the form of beta and gamma rays has been considered for sludge disinfection. Both beta and gamma rays induce secondary ionizations in sludge as they penetrate, thereby inactivating the pathogens and producing oxidizing and reducing compounds which attack and destroy them.

Conditioning

Conditioning sludge helps improve its dewatering properties. In addition, conditioning sometimes disinfects wastewater solids, removes odors, alters the physical characteristics of the solids, destroys solids on a limited basis, as well as improving solids recovery. Conditioning can be accomplished by biological, chemical and/or physical means.

Lime and ferric chloride are popular inorganic chemicals used in the sludge conditioning process. Another chemical which has been used in this application on a limited basis is ferrous sulfate. Inorganic chemical addition is used in dewatering processes such as mechanical sludge dewatering and vacuum filtration. Ferric chloride addition to sludge produces positively charged iron ions which react with negatively charged sludge ions, neutralizing them and allowing them to form larger aggregates which settle easily. Reaction of ferric chloride with bicarbonate alkalinity leads to the formation of hydroxides which act as flocculants, further improving sludge settleability. Hydrated lime is added to sludge together with ferric ion salts. Lime has a slight dehydration effect. However, its main purpose is to raise pH, reduce odors and disinfect the sludge. Calcium carbonate formed by the addition of lime to sludge provides a granular structure which increases the porosity of the sludge and reduces sludge compressibility.

Polyelectrolytes are long chain molecules which are synthesized by many chemical manufacturers today. Polyelectrolytes are used in sludge conditioning frequently. These chemicals are used to desorb water from the surface of solid particles, neutralize the charge of the solid particles, and act as a bridge between particles, thereby facilitating agglomerization of small particulates.

Sludge conditioning can also be done without the use of chemicals. One example is the use of power plant fly ash to improve sludge dewatering qualities. Another example is use of pulverized coal for conditioning of wastewater sludge. Pulverized coal is an excellent conditioning step when the treated sludge is to be used for its fuel value in a combustion process such as incineration, since the coal helps to increase the fuel value of the sludge.

Other processes used for sludge conditioning include thermal processing, elutriation, the freeze-thaw process as well as screening and grinding. While thermal processing produces a sludge with a high heating value well-suited for use in combustion processes, it also produces an odorous gas sidestream and has high capital costs associated with it. Elutriation is a process by which anaerobically digested sludge is physically washed to dilute bicarbonate alkalinity and reduce acidic metal salt demand. Thus, less metallic salts are required to neutralize negative charges in the sludge, which facilitate solids agglomeration and subsequent settling. The freeze-thaw process improves sludge drainage qualities, and is also considered a conditioning process. Screening and grinding of sludge reduces the amount of large particles in the cake, making the cake more uniform and easier to pump.

Dewatering

Dewatering of wastewater sludge reduces its volume and results in a non-fluid material more suitable for ultimate disposal. Dewatering decreases the volume of water in the sludge to a far greater extent. Though popularity has increased, centrifuges are still not as widely used as natural and gravity type dewatering methods. Two types of centrifuges which are available for sludge treatment are the imperforate basket and the scroll-type decanter.

Filtration is a process by which solids are removed from a liquid stream by passing the liquid through a bed of porous media. Surface filtration uses driving forces such as a vacuum, pressurization, gravity, or even centrifugal force to result in a pressure drop through the filter media with subsequent settling of solids on the filter bed. Surface filters can be defined as filters in which solids are deposited in the form of a cake on the upstream side of a relatively thin filter medium.

Developments in the filtration field have led to the design and production of many different types of filters. Vacuum filters employ atmospheric pressure downstream of the filter media to drive the liquid through the porous media resulting in the deposition of solids on the filter bed. Refer to **Figure 5** for an example. Belt filter presses use two belts to mechanically compress the sludge cake thereby forcing water out and reducing its volume by as much as 50 percent. Pressure filters use increased pressure as a result of sludge addition to the pressure vessel in order to force water from the sludge cake. Sludge is added to the vessel until a pressure close to the design pressure of the vessel is reached. Upon reaching this pressure, the vessel is opened and the dewatered sludge cake is removed.

Mechanical filters such as the screw and roll press are also becoming popular for sludge dewatering. In a screw and roll press, the sludge is pumped inside a screen surrounding a screw. The sludge is deposited against the screen wall by the continual rotation of the screw. The screw and roll press is best used to dewater primary sludges.

Other filtration methods available are the dual cell gravity (DCG) filter and tube filters. The DCG filter is most efficient when used to dewater waste activated sludge. The tube filter has been widely used in industrial applications, however, there is no record of their use in municipal sludge dewatering.

Heat Drying

Drying of sludge by the use of heat has not gained wide popularity in the United States.

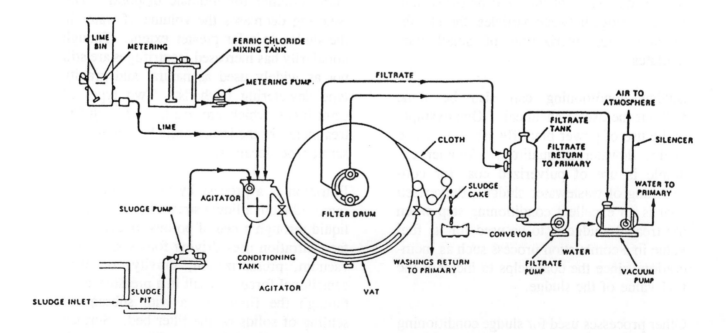

Figure 5.   Rotary vacuum filter system

Although this process has been used for many years, it is generally employed in the treatment of sludges to be used for soil conditioning. The high cost of fuel needed to dry the sludge, combined with the sparse market for the heat dried sludge have made this treatment process an unpopular alternative.

Heating drying of sludge is carried out at a temperature too low to destroy organics present in the cake. In addition, water vapor containing contaminants from the sludge poses a potential air pollution control problem. Various types of heat drying operations have been designed. One of the more popular systems in both industrial and municipal applications is a multiple-effect evaporation system.

Thermal Processing

Treatment of sludges using thermal methods has become increasingly popular. Thermal destruction provides maximum volume/weight reduction. High operating temperatures result in detoxification of the pathogens present in the sludge. Thermal destruction processes use the heat content inherently present in the sludge. Disadvantages of thermal processes include high capital and operating costs, operational problems and air pollution control system requirements. Various types of technologies, both proven and unproven are available for thermal destruction of wastewater sludges. Some of the more popular processes for this application include incineration, fluid bed combustion and multiple hearth furnace combustion.

UTILIZATION OF SLUDGE

Treated wastewater sludge can be disposed of or utilized for its valuable properties.

Utilization is the process by which the sludge itself or its by-products are put to beneficial use. Depending on the properties of the sludge or sludge by-products, they can be added to soil for nutritional value or stabilization, incinerated for their heat content and even used for chemical value; as animal feed, etc.

Land Application

Many forms of wastewater sludge contain valuable nutrients which are beneficial to the growth of farmland crops. Several communities in the United States today utilize sludge for various purposes through land application. Land application of wastewater sludges has advantages, particularly because of its nutritional value. However, the application process must be carefully monitored for parameters such as odors, as well as contamination of groundwater and surface waters. In addition to being applied to croplands, a sludge has been used as a stabilizing material in foundations of buildings, walls and dikes. Potential utilization of sludge as a stabilizer eliminates the need for its ultimate disposal and can put a material once considered useless to good use.

Energy Recovery

Sludge materials often have high heat contents and can be burned to generate power. In addition to the energy supplied by the solid sludge material, energy is also recovered in the form of methane gas produced during anaerobic digestion. Called biogas, because it is generated by the conversion of organic material to methane in the presence of microorganisms, this gas has a high heating value (600 BTU/SCF) and has made anaerobic digestion a viable alternative. Biogas is also produced by the action of anaerobic bacteria in sludge land-fills and has been recovered as an energy source. The problem associated with the use of biogas and sludge as energy alternatives are mainly air pollution problems. While some sludges contain a great deal of highly combustible materials, they also contain contaminants such as H_2S, which upon burning, form air pollutants like SO_x. Therefore, in designing boilers and incinerators to burn sludges or biogas, the engineer must provide flue gas cleaning equipment to eliminate air contaminants.

Resource Recovery

In addition to being utilized for land application, land stabilization or energy recovery, some wastewater sludges contain materials which can be reused for their chemical or animal feed values. Some recovery processes are very expensive, thus an economic study must be conducted on a case-by-case basis.

Chemicals recovered from wastewater sludge can be used in the wastewater treatment process or sold. Recovery is more feasible when the sludge contains relatively high concentrations of the particular material whose recovery is desired. When very low concentrations of the chemical are present, recovery may not be worthwhile and the sludge will be disposed of in a conventional manner unless removal becomes a regulatory requirement such as in the case of heavy metals or toxic organics. Some wastewater sludges have been used for animal feeds. However, certain sludges contain high levels of elements, such as cadmium, which are toxic to animals. Before use as an animal feed material careful chemical analysis of the sludge should be conducted.

Sludges have been used as construction materials. Due to its cement-like properties, sludges chosen for this application are mixed as a supplement to the main building material. By doing so the building material becomes stronger and the sludge disposal problem is eliminated.

ULTIMATE DISPOSAL

Wastewater sludges which have no physical or chemical constituents worth recovering must be ultimately disposed of. The most common means of disposal has been the sanitary landfill. Sanitary landfilling has been used for quite a number of years and has been successful in sludge disposal applications where viable.

Sludge landfilling refers to a process by which wastewater solids such as grit, scum and treated sludge are disposed of in a segregated landfill. No other materials are landfilled with the wastewater solids. Sludge landfills fall into two basic categories, trenches and area fills. Trenches are landfill areas which are dug below the original ground surface and wastewater solids are deposited. Narrow trenches are those which are less than 10 feet in width, while wide trenches are those which are greater than 10 feet in width. Area fill is a process by which sludge is mixed with the soil and placed above the ground. The sludge/soil mixture can either be put in place in the form of a mound, called an area fill mound, or it can be spread evenly over the existing grade and is therefore called an area fill layer. Landfilling of wastewater sludge requires prevention of leaching of contaminants into groundwater supplies. Such prevention can be obtained by providing an artificial liner such as clay with or without an underdrain system. When a wastewater sludge proves to be hazardous, based on RCRA leachate criteria, it must be disposed of in an EPA approved landfill. If an EPA approved landfill is not present on-site, the sludge must be transported to a landfill having the required EPA permits for disposal of hazardous waste.

Co-disposal is the landfilling of wastewater sludge together with other refuse. Such disposal often presents complicated leaching problems since heterogenous materials are contained in the landfill. Careful characterization of the sludge is required in order to determine whether it is hazardous or nonhazardous. Leachate treatment facilities must be designed to treat the leachate from these very complex landfills.

Advantages of disposing of sludge in a sanitary landfill include the fact that it is a well-known and extensively used process; biogas (methane gas) which can be used for combustion is produced in the inner layers of the landfill where anaerobic conditions exist; and landfilling is relatively inexpensive when compared to recovery or combustion disposal processes. The main disadvantage of landfill disposal is the unavailability of land and regulatory restrictions. As more and more land is used for building and construction as well as for landfilling, less is available for landfilling in the future.

Recent developments in the utilization of wastewater sludges combined with improvements in sludge combustion processes have helped to lessen the amount of sludge required to be disposed of in sanitary landfills.

SLUDGE REDUCTION

Methods, Functions and Occurrences

Sludge reduction processes are generally thermal ones and provide a major reduction in the sludge solids. Although heat drying is not truly a reduction process, it occupies the same relative position in the sequence of sludge processing as major reduction processes. It is included in this chapter for purposes of simplification. Established and experimental sludge reduction processes are listed and categorized in the table below.

REDUCTION PROCESSES

Reduction Process	Pretreatment Required	Additional Processing Requirements
Established Processes		
Incineration	Thickening and Dewatering	Landfill ash
Wet Air Oxidation	Thickening	Treat cooking liquor, Landfill ash
Heat Drying	Thickening and Dewatering	Use dried sludge as soil conditioner
Experimental Processes		
Pyrolysis	Thickening	Utilize by-products of gas, carbon, steam. Dispose of residue
Incineration/Power or steam generation	Thickening and Dewatering	Landfill ash

FIXATION OF SLUDGES

Fixation refers to a broad class of treatment processes that physically or chemically reduce the mobility of hazardous constituents in waste. Other terms that are sometimes used synonymously for fixation are solidification and stabilization. The stabilization treatment system consists of a feed system, a tank equipped with mixing equipment, and a cure area.

This technology has wide application to the California List of metal wastes. In most instances, the technology is used where wastes of interest already have a significant percentage of solids, e.g. metal precipitates in treatment sludges. Stabilization can be applied to wastewater.

The underlying principle of stabilization is the binding of constituents of concern into a solid resistant to leaching. The mechanism by which this occurs depends upon the type of stabilization process. Two of the most common systems are lime/pozzolan based processes, and Portland cement-based processes.

In Portland cement systems, the waste is mixed in a slurry with anyhydrous cement powder, water and frequently, pozzolanic additives. The cement powder is a mixture of powdered oxides of calcium, silica, aluminum and iron produced by kiln burning materials rich in calcium and silica at high temperatures. The major mechanism of stabilization in this system is the formation of hydration products from silicate compounds and water. A calcium silicate hydrate gel forms. This gel then swells and forms the cement matrix composed of interlocking silicate fibrils. At the same time, constituents present in the waste slurry, e.g., hydroxides of calcium and various heavy metals form the interactives of the cement matrix. Metal ions may also be incorporated into the crystal structure of the cement matrix itself. A rigid mass results from the interlocking fibrils and other components during setting and curing.

Lime/pozzolan processes use the finely divided, noncrystalline silica in pozzolanic material (e.g., flyash) and the calcium in lime to produce a concrete-like solid of calcium silicate and alumina hydrates. Waste containment is achieved by entrapping the waste in this pozzolan concrete matrix. In actual operation, the waste, water and a selected pozzolanic material are mixed to a pasty consistency. Hydrated lime is blended into the mixture and the resulting moist material is packed or compressed into a mold and cured over a sufficient time interval.

The level of performance for stabilization processes is measured by the amount of constituents that can be leached from the stabilized material. There are two techniques currently recognized by EPA as measures of leachability. The first is the Extraction Procedure (EP) Toxicity Test (40CFR261); the second is the Toxicity Characteristic Leaching Procedure (TCLP).

CONCLUSIONS

A large number of methods are available for the treatment and disposal of sludges. The choice of sludge treatment depends on the final disposition of the treated sludge as well as on regulatory and economic considerations. The ultimate disposal method for sludge is a function of its characteristics. Where it is economical to do so, valuable material in the form of nutrients, energy, and chemicals can be recovered.

Of the treatment technologies available, some have been used extensively, while others are relatively new and innovative. Simple gravity thickening and dewatering has been employed historically for the treatment of wastewater sludges. Anaerobic digestion has been used for stabilization of wastewater sludge, however, recent improvements in the process have led to the generation of methane gas or biogas; this has resulted in an increase in the popularity of this process for sludge stabilization.

Improvements in combustion processes such as incinerators and boilers have led to the recovery of energy from sludge. In addition, land application of sludge is a means of disposing of treated sludge while putting it to good use. Some treated sludges have been used for animal feed. The concentration of toxic trace elements such as cadmium in many sludges has limited this application. Advances in the sanitary landfilling procedure has led to both safer and more secure landfills. Development of impermeable artificial liners together with extensive leachate collection systems has helped improve ground water quality while relieving the worries of hazardous contaminant exposure.

CHAPTER 6

RECOVERY SYSTEMS FROM WASTES DISPOSAL

INTRODUCTION AND OVERVIEW

Modern society today is the product of constantly advancing technology. Technological progress is responsible for our high overall living standard and also for pollution and waste. Fortunately, technology can also be used to solve these problems.

Concerns for the environment are not limited to detrimental effects of pollution, but also include recovery and utilization of resources now recognized as finite, namely fuels and energy. Savings can be realized by many industries, institutions and governments through the potential of converting their own solid wastes into energy and resources. Recovery possibilities offer new opportunities when it becomes apparent that it is bad economics to pay someone to haul wastes away for disposal. Possibilities are:

- contribution to energy supply,

- relief from many environmental problems associated with wastes disposal.

Industrial application of incineration technology involving energy recovery has historically designed for hazardous material and wastes destruction to be achieved by fuel fired boilers operating usually at reduced firing rates or by fired boilers designed for a specific hazardous waste. Reduce rate firing has the disadvantage of reduced steam/energy production. Hazardous wastes firing other than the design waste runs the risk of inefficient destruction and regulatory violation. More recent concepts incorporate high heat recovery efficiency as well as desired destruction efficiency.

LIQUID WASTE FUEL COMBUSTION

In recent years, combustion of certain classifications of waste oils and spent solvents in industrial boilers has increased. Increases in fuel prices in the mid 1970's provided a strong incentive toward waste fuel combustion with heat recovery in the form of steam generation. Economic viability and interest in burning hazardous waste liquids in conjunction with fossil fuels also began to increase rapidly in 1976 when the Resource Conservation and Recovery Act (RCRA) was passed by Congress. RCRA's stringent prohibition of landfilling hazardous waste oils and solvents, dramatically increased waste disposal costs. Because many types of waste oils and solvent exhibit properties of potentially good fuels, combustion of these wastes provides the double incentive of reduced steam generation costs, and solving an otherwise expensive waste disposal problem. Waste oil and solvent combustion in boilers also supports the philosophy of resource recovery.

The term "waste oil" refers to used motor vehicle crankcase oils and spent machinery lubricating oils. The quantity of waste oil generated in the United States is estimated to be 1.1×10^9 gallons per year of which approximately 40 percent is currently burned as fuel. Additives include barium, magnesium, zinc, sulfur, nitrogen, calcium and phosphorus. During use, lubricating oils may also become contaminated from both internal and external sources. For example, when leaded gasoline is used in an automotive engine, the crankcase oil becomes contaminated with lead via the piston rings and cylinder walls (external). Lead alloys are also used as bearing material inside many engine crankcases (internal). The moving parts of machinery or engines also wear, causing internal oil contamination with metals such as iron, chromium, nickel, molybdenum, aluminum, zinc and magnesium. Environmental impacts of the disposal of waste oils has been an area of concern. Several studies conducted by State and Federal agencies have documented the presence of contaminants such as chlorinated hydrocarbons and the above mentioned metals in samples of used motor oils.

The term "spent solvent" refers to a broad classification of waste liquid hydrocarbons. These solvents are used by a large group of industries. They are used in chemical or pharmaceutical processes including reactions, extractions, degreasing or cleaning operations. Many spent or contaminated solvents are recovered by filtration and distillation or other means to purify the material. However, often the solvent may become contaminated with organic residues or other solvents that render it difficult or too expensive to recover. Generally only non-halogenated solvents are suitable as waste fuels for industrial boilers. Concentrated halogenated solvents are generally poor fuels and their high halogen content render them illegal to burn without acid gas scrubbers. **Table 1** shows examples of typical solvents utilized by the pharmaceutical industry for example which may be suitable boiler waste fuels.

Types of Industrial Boilers - It has been estimated that at least 25,000 boilers designed to fire fuel are partially fired by hazardous wastes. Most of the industrial boilers currently burning waste fuels were originally designed for natural gas and or fuel oil combustion. Because of competitive market conditions fuel fired boilers are designed for maximum volumetric heat release. The most common boiler designs are firetube and watertube.

Firetube Boilers - The name firetube boiler is derived from the fact that in boilers of this type heat is transferred from hot combustion products flowing inside tubes to the water surrounding them. Fuel combustion takes place in a cylindrical furnace within the boiler shell. Firetubes run the length of the shell at the sides of, and above, the internal furnace. Gas from the furnace reverses direction in a chamber at the rear and travels forward through the tubes to the front of the boiler. Typical fire tube boiler capacities range from 10,000 to 30,000 lb/hr steam.

Watertube Boilers - Watertube boilers contain steel tubing (containing water) throughout the combustion chamber. Heat from the combustion products is transferred from the path of the flue gas into the adjacent watertubes. The walls of a water tube boiler are often lined with water tubes to assure minimum heat loss from the boiler shelf. Typical watertube boiler capacities range from 20,000 to 250,000 lb/hr steam.

TABLE 1.

TYPICAL SOLVENTS SUITABLE FOR WASTE FUEL BLENDING

Ethyl Acetate	Cyclohexane
Acetone	Diethylaniline
Methanol	Diethylamine
Ethyl Ether	Methyl Vinyl Ketone
Toluene	Butanol
Hexane	Dimethoxy Propane
Heptane	Acetic Acid
Isobutyraldehyde	Tetrahydrofuran
Methyl Formate	Methyl Benzyl Ether
Ethanol	Benzyl Alcohol
Propionic Acid	Dibenzyl Ether
Propionic Anhydride	Acetic Anhydride
Methyl Ethyl Ketone	Cyclohexyl Acetate
Dimethylaniline	Triethylamine
Isopropanol	Cyclohexylethylamine
Dimethylformamide	Cyclohexenylethylamine
Tetrahydrofuran	Benzaldehyde
Butenediol	Benzylamine
Xylene	Acetonitrile
Pyridine	Butenol
Ethyl Butenol	Butyl Acetate
Methyl Acetate	Methyl Isobutyl Ketone
3-Hexynol	Isopropyl Acetate
Aniline	Ethyl Benzene
Alcohol 2B	Dimethyl Malonate
Isopropyl Ether	Monobenzylamine

*   It should be noted that specific states may not allow carcinogenic or suspect carcinogenic matter to be burned in industrial boilers.

Retrofitting an Industrial Boiler for Liquid Waste Fuels

The capital investment to convert an existing industrial boiler burning either natural gas or conventional fuel oil to waste fuel operation can be divided into five major components:

- Liquid waste storage and feed systems;

- Boiler modifications to accommodate the waste fuels;

- Engineering and permitting fees necessary to properly license the boiler with the appropriate state environmental agency;

- Air pollution instrumentation including continuous stack monitoring and recording devices, and

- Air pollution control devices (if required).

Liquid Waste Storage and Feed Systems - In many cases liquid wastes are transported from a process area to the boiler via tank truck and pumped into a storage tank. If the boiler is reasonably close to the waste generation area, direct pumping by a permanent pipeline is also feasible. A separate settling tank is good engineering practice if the waste contains settleable solids.

If the waste is delivered via tank truck, a single 8,000 gallon storage tank is usually sufficient. This capacity will handle a typical 6,000 - gallon tank truck load, and will provide a buffer storage capacity for continuous boiler operation. In some cases, a larger storage capacity may be desirable should blending or equalization be desired.

If the waste is continually pumped directly from the generation area, a large storage tank will permit continued operation on either end if waste generation or boiler operation are intermittent. Storage tanks containing flammable liquid fuels should ideally be stored below ground in double-walled tanks. The free space between the tank shells should be monitored for liquid leaks. Provisions must also be made to inert the tanks with nitrogen and flame arrestors must be installed on the tank vents. It may be necessary to provide agitators on the tanks if the waste may physically separate into different phase layers.

Wastes are normally pumped from the storage tank directly to the boiler through strainers and filters. Two feed pumps are often desired in the event one pump breaks down. Transfer lines should be double-walled design with the free space between the pipes monitored for leaks.

Boiler Modifications - For most single burner firetube or watertube boilers originally designed for combination gas/oil firing, the only necessary modifications are oil gun replumbing for liquid waste injection and minor control system changes. With these modifications, liquid waste can be fired through the center fuel gun, while natural gas or fuel oil is fired through the outer ring to sustain the desired steam load. Under reduced load conditions, waste firing would be terminated first and either fuel oil or gas alone would be burned.

A second alternative for single burner boilers is to install a mixing system for fuel oil and waste upstream from the nozzle. The boiler would then be operating on a consistent mix of fuel oil and waste fired through the existing oil gun. However, this is usually feasible only if:

1. the waste is of consistent composition, and

2. the waste and fuel oil are miscible and compatible with respect to volatility and viscosity.

Otherwise, serious combustion control problems may occur.

The third alternative is replacement of the entire burner with a model specifically designed to accommodate liquid waste, cofiring with oil or gas. This is the preferred choice with respect to long-term operating efficiency and maintenance considerations.

Chemical Analysis and Permitting Fees - Based on experience with testing laboratories, the following fees may be most likely incurred in the boiler permitting process but may vary depending on local regulations.

- Initial isokenetic stack sampling of boiler $6000 - $11,000

- Atomic absorption metals analysis of waste fuel - $25 per each metal analysis

- Halogen analysis - $75/sample

- Btu value of fuel - $50/sample

Typical air permit application fees are currently about $250 per boiler.

Safety Consideration

Typical spent solvents are all flammable liquids. This means they have closed cup flash points below 100°F.

Conventional fuel oils generally have flash points of over 130°F and are classified as combustible liquids. Handling, storage and distribution systems for industrial boilers being converted to flammable waste combustion must be safety designed. In the event a transfer line breaks inside the boiler room, flammable vapors could accumulate and explode upon contact with an ignition source. It is good engineering practice to locate all flammable liquid storage tanks and feed pumps outside in an area equipped with spill control provisions. Transfer lines should be double-walled construction.

Another key safety area to consider is generally termed "boiler flameout." This is a potentially hazardous condition in which a slug of residue or water enters the boiler burner nozzle and causes a temporary flame extinguishment. After the slug passes, raw fuel is aspirated into the boiler chamber where it may accumulate and then explosively ignite. Generally, boilers that cofire waste fuels and conventional fuels simultaneously through two separate nozzles do not exhibit this explosive flame-out condition. However, in boilers firing only one nozzle at a time, this condition must be carefully addressed and controls installed to prevent this occurrence.

AIR EMISSIONS

Emission Characteristics - The combustion of specific waste oils and solvents in industrial boilers must be closely examined from an air emissions standpoint. Until recently, very little documentation regarding the extent to which chemical contaminants or hydrocarbons constituents in waste fuels are destroyed or altered during boiler combustion was available.

In 1984, the United States Environmental Protection Agency (USEPA) conducted a comprehensive study of the air emission impacts from the disposal of waste oils by combination in commercial boilers. In this study, actual stock tests were conducted on boilers in the size range of 0.4 to 25 million Btu/hr. Seven boilers were selected for testing in the program. The units were chosen to provide a representative cross section of the types and sizes of boilers previously described. A 4,000 gallon supply of used automotive oil was obtained and served as a consistent supply of waste fuel for the program. Some of the stock oil was spiked with measured amounts of selected organic compounds which are typically found in waste fuels. The selected organic compounds were chloroform, 1,1,1 - trichloro-naphthalene, 2,4,5 trichlorophenol and chlorotoluene. Stack testing was conducted at each of the sites to determine atmospheric emissions of particulates (principally lead) inorganic compounds (principally HCL) and volatile organic compounds. The destruction efficiencies for each of the spiked components were also determined.

From the results of this study, it is possible to make some general conclusions concerning the combustion of waste automotive crankcase oil in boilers of this size range:

1. The emissions of lead and other metals including arsenic, cadmium, are significant and of immediate concern. Most of the lead emissions are submicron in size, and they are readily inhalable. Material balance calculations indicate that 50 to 60 percent of the lead introduced into a boiler exits from the system via the stack. Analysis of the ash collected in the firebox indicates lead levels of up to 2 percent. This provided an accounting mass balance of only 65 percent of the lead consumed. The National Ambient Air Quality Standard (NAAQS) for lead may possibly be violated in the impact areas surrounding an industrial boiler combusting pure automotive waste oils.

2. It is possible to achieve hydrocarbon combustion efficiencies greater than 99.9 percent for industrial boilers firing waste oils. For the spiked halogenated organic compounds typically found in trace quantities in waste oils, destruction efficiencies greater than 99.9 percent are obtainable.

3. Particulate emissions from the six boilers tested range from 0.97 to 1.2 lbs/hr (0.34 lb/million Btu heat input). This is significantly higher than the EPA's own emission factors of 0.09 lb/million Btu for commercial boilers firing residual oil. However, the higher value is consistent with the much higher ash content of waste oil, which can range from 0.15 percent to 1.5 percent.

4. In the boilers tested that were above one million Btu/hr capacity, there was no apparent correlation between boiler size or firing method and hydrocarbon destruction efficiency.

5. Polychlorinated dioxin (PCDD) and chlorinated dibenzofuran (PCDF) and species were detected in 60 percent of the boiler stack samples. The concentrations of these toxic particulate contaminants ranged from 7 to 470 parts per trillion (ppt). Tests were also completed on samples of

the waste fuel to determine PCDD and PCDF levels prior to combustion. No dioxin or dibenzofuran compounds were detected in any of the oil samples. Therefore, dioxin and dibenzofuran found in the boiler stack sampling were probably formed during the combustion process. The fly ash deposited inside the boilers may contain parts per billion levels of chlorinated dibenzofuran and dioxin compounds. This has the potential for being classified as hazardous on this basis, and may be subject to stringent RCRA regulations for disposal. The whole dioxin issue is not specific to industrial boilers burning waste fuels. Dioxins are an important environmental issue in assessing the impacts of garbage burning resource recovery plants, and hazardous waste incinerators. It has been said that tetrachloridebenzodoxin (TCDD) is the most hazardous chemical ever made by mankind. The extent to which the family of PCDD and PCDF compounds pose a hazard at the low levels found in the stack gas of industrial boilers burning waste fuels is undetermined.

6. Halogen based acid gas emissions from the boilers tested were significant. Halogen elements include chlorine, bromine and flourine. These elements when present in waste fuels usually form hydrogen chloride, hydrogen bromide or hydrogen fluoride in the flue gas when combusted. These compounds are all acid gases that present both localized and long-range transport air pollution problems.

Control of "acid rain" is a key issue on both the State and Federal regulatory levels. Combustion of waste fuels with high halogen levels in industrial boilers without acid gas scrubbers is illegal.

Instrumentation - The quantity of oxygen in the flue gas is a good indicator of the status of a boiler's combustion process. The presence of oxygen indicates that excess air is being introduced. In boiler operation, it is necessary to provide a system to allow automatic proportioning of the quantity of air to the quantity of fuel. The three types of combustion controllers used in boilers include: fuel-flow, air-flow, and gas-flow analysis. The type of boiler and the fuel properties must be examined to determine the best guide for a particular unit. A combination is sometimes incorporated in the instrumentation system.

To license and operate an industrial boiler to burn waste fuel in New Jersey, for example, a boiler must be equipped with a flue gas analysis system. Installation of a continuous oxygen analyzer and either a continuous carbon monoxide or total hydrocarbon analyzer in the boiler stack is required. Each analyzer must be hooked up to a permanent recording device. The make, model and specific types of monitors and recorders must be approved by the regulatory agency. Continuous monitoring devices are expensive and can be a substantial segment of the total retrofit costs. A major product of inefficient combustion, when burning any waste fuel, is carbon monoxide. When fuel is burned without sufficient oxygen or without adequate turbulence mixing, large quantities of carbon monoxide are emitted along with unburned quantities of fuel. Some carbon monoxide is always present in the boiler combustion process, no matter what the level of excess air. With high levels of excess air CO typically ranges from 5-20 parts per million (ppm). If the excess air is reduced and the oxygen level approaches stoichiometric incineration to the waste fuel, the CO level will suddenly start to increase.

Figure 1 illustrates the relationship between excess or unburned CO in the flue gas. Most CO analyses on boilers are infrared (IR) detection devices and use a high resolution absorption spectroscopy technique that measures from 0 - 1,000 ppm CO. An infrared light beam is sent across the stack from an infrared light source on one side, to a detector on the other. In front of the detector is a slit band filter that allows only the portion of the IR spectrum that is sensitive to CO to pass. The instrument only receives the residual IR present that hasn't been absorbed by the CO in the stack. Thus the residual IR is proportional to the CO level in the stack.

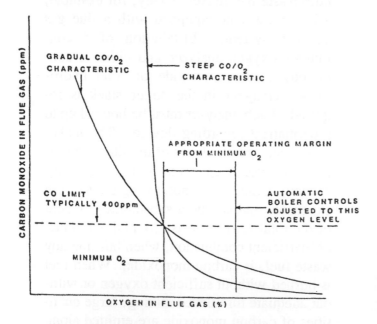

Figure 1. **Relationships between CO-CO$_2$ in the flue gas (note: the fuel/air ratio on the dotted line is the optimum operating condition)**

A common type of oxygen (O$_2$) sensor used in boilers is a zirconia cell detector. The detector measures percent O$_2$ by reading the voltage across a heated electrochemical cell, consisting of a small zirconia disc. Both sides of the disc are coated with porous metal electrodes. When operated at a temperature maintained by a Temperature Controller, the output voltage (millivolts) of the cell is given by the following Nernst equation:

$$EMF = KT \log 10 \ P_1 P_2 + C$$

P$_1$ is the partial pressure of the oxygen in the reference gas on the other side, and P$_2$ is the partial pressure of the oxygen in the measured gas on one side of the cell. Normal air from a clean, dry, instrument air supply (20.95% oxygen) is used as the reference gas. T is the absolute temperature; C is the cell constant and K is an arithmetic constant.

When the cell is at operating temperature and there are unequal oxygen concentrations across the cell, oxygen ions will travel from the high partial pressure of oxygen to the low partial pressure side of the cell. Because the size of the output signal is proportional to the log of the inverse of the sample oxygen concentration of the sample, the gas decreases. This characteristic enables the O$_2$ analyzer to provide evident sensitivity at low O$_2$ concentrations. Zirconia cell oxygen detectors measure oxygen concentration in the presence of all combustion products (including water). It may therefore be considered as an analysis on a "wet" basis. In comparison, other methods, such as the Orsat apparatus, provide an analysis on a "dry" gas basis. "Wet" analysis will, in general, indicate a lower percentage of oxygen. The difference will be directly related to the water vapor content of the sampled gas stream.

Air Emission Control Devices - Industrial boilers that fire either conventional or waste fuels are generally not equipped with any air pollution control devices (APCD's). However, with todays rising cost of hazardous waste disposal of many spent solvents, it is becoming increasingly attractive and cost effective to consider adding an APCD to allow boiler combustion of a waste fuel stream that may not meet the regulatory requirement for uncontrolled boiler combustion. The two criteria pollutants that may present a need for control are:

1. particulates, and

2. hydrogen chloride (HCl).

Particulate emissions stem from the ash content of the fuel and HCl emissions from the fuel's halogen content.

FEDERAL REGULATORY REQUIREMENTS

On November 29, 1985, the final US Federal rules pertaining to waste fuel combustion in boilers was published in the Federal Register. Prior to this regulation, the USEPA received adverse publicity for loopholes in RCRA that exempted regulation of boilers burning waste fuels. Under this regulation, the USEPA places both stringent controls on actual burning and administrative controls on companies who market and burn hazardous waste and used oil for energy recovery. The controls on burning prohibit non-industrial boilers from burning hazardous waste and used oils.

Refuse Energy Recovery Technologies

Refuse energy recovery technologies may be grouped in a number of classifications:

- Direct combustion of unprocessed municipal refuse (Mass Burning).

- Preparation of refuse-derived fuel and direct combustion.

- Pyrolysis/Gasification.

- Cogeneration.

- Biological processes.

- Direct Combustion of MSW

Most of the energy recovery facilities in the US burn unprocessed refuse directly. There are many modular incinerators and waterwall boilers in operation.

Refuse Derived Fuel (RDF)

Refuse Derived Fuel (RDF) is prepared by several unit processes, including shredding, screening, magnetic separation and air classification. RDF can be burned in dedicated semisuspension-fired boilers, fluidized bed boilers and a supplemental fuel with coal or oil in utility boilers. A typical fluidized bed combustion system is illustrated in **Figure 2**.

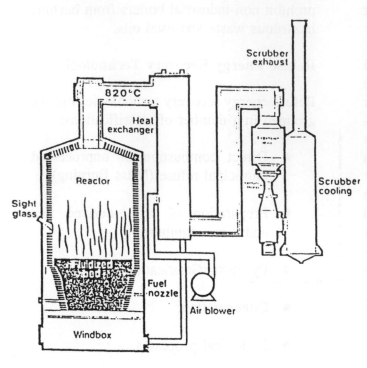

Figure 2. Fluidized bed combustion with energy recovery

Pyrolysis/Gasification

A number of pyrolysis and gasification processes have been developed for municipal refuse, but none are operating in the US at the present time, due to a variety of technical and economic problems. These include the Union Carbide PUROX and Andco-Torax fixed bed processes, Japanese fluidized bed technology and Occidental Flash Pyrolysis Entrained Bed Process. Andco Torax plants that have been built include one at Walt Disney World in Florida.

Cogeneration

Cogeneration is the simultaneous utilization of heat and power from a single thermodynamic cycle. In a typical steam-topping-cycle cogeneration system, fuel is burned in a boiler to turn the water to steam which is piped to a turbine to drive a generator. The electricity produced can be used by the plant or sold. Since the steam from the turbine retains much of its energy, it can be used again for heating or other applications. **Figures 3** through **5** illustrate components of cogeneration operations.

Biological Processes

Of the biological processes, only methane recovery from landfills is commercial and landfill gas recovery projects were in operation in California and New York. Enzymatic hydrolysis can convert cellulosic wastes to ethanol in a fermentation reactor, and several laboratory scale plants have been built to convert municipal refuse to alcohol. This latter process continues to be studied. Refer to **Figure 6**.

Mass Burning of Municipal Waste with Energy Recovery

Mass burning in a refractory or water wall furnace with a heat recovery boiler has been a proven system for over 25 years in Europe and has had some record of success in the US. More than two dozen facilities have at one time or another generated energy from MSW (municipal solid waste) by waterwall furnaces in the US. Some of these include the following.

Pinellas County, Florida - The plant was completed in May, 1983 and operated after experiencing ordinary shakedown problems. An expansion was begun to accommodate 3000 TPD.

Chicago (Northwest) - The plant is processing 1100 TPD after completion of a major maintenance program. Steam is supplied to Brach Candy Company.

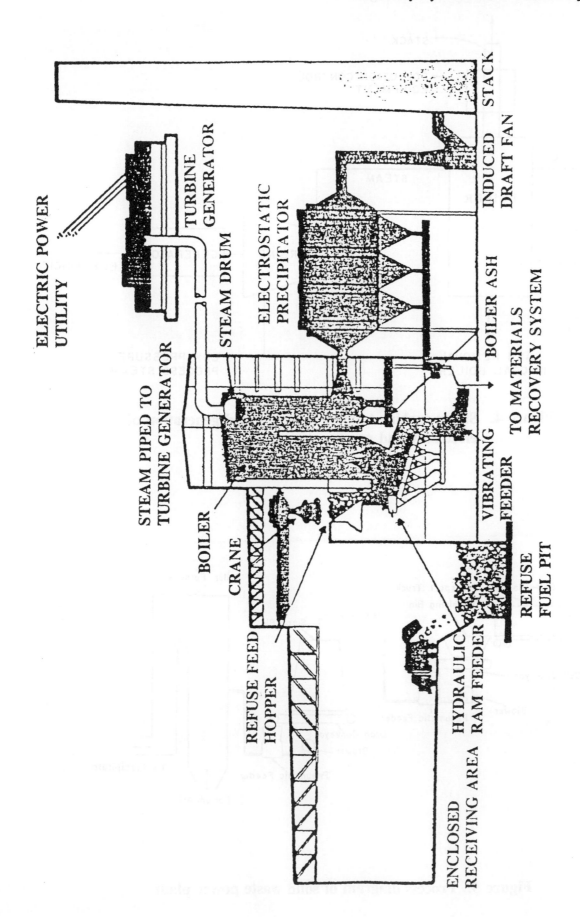

ELECTRIC POWER UTILITY

TURBINE GENERATOR

STEAM DRUM

ELECTROSTATIC PRECIPITATOR

STACK

INDUCED DRAFT FAN

STEAM PIPED TO TURBINE GENERATOR

BOILER ASH

TO MATERIALS RECOVERY SYSTEM

BOILER

CRANE

VIBRATING FEEDER

REFUSE FEED HOPPER

REFUSE FUEL PIT

ENCLOSED RECEIVING AREA

HYDRAULIC RAM FEEDER

Figure 3. Sectional view of a refuse-to-energy system

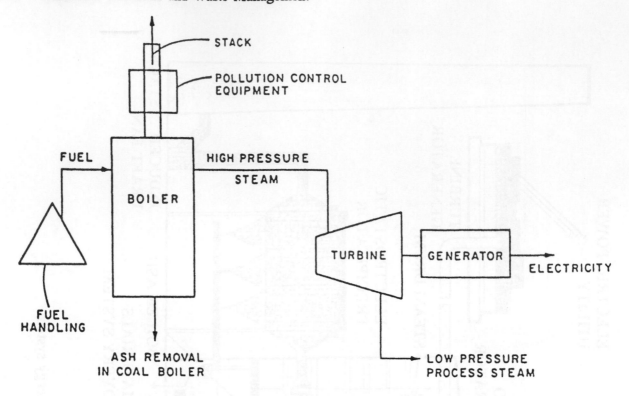

Figure 4. Steam turbine topping for combined generation of electricity and process steam

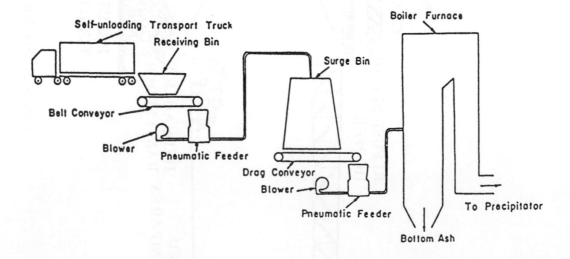

Figure 5. Process diagram of solid waste power plant

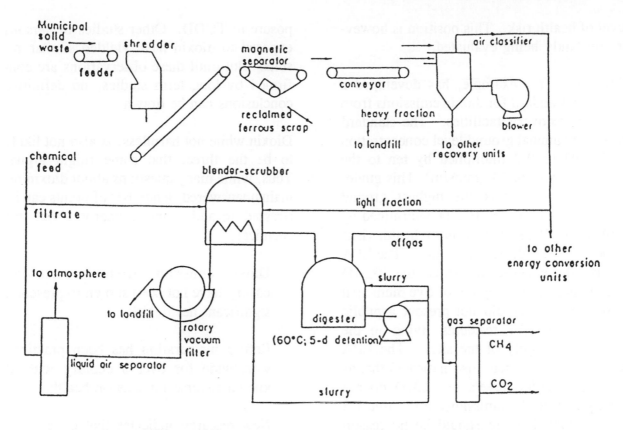

Figure 6.   Solid waste conversion process to produce methane

Saugus, Massachusetts - This facility was to supply steam to Bethlehem Steel; Plant underwent major overhaul. Disposal of sludge and cogeneration of electricity has been planned.

Nashville, Tennessee - This plant has processed 440 TPD during one year. Under planned expansion, electricity was to be produced and sold to Tennessee Valley Authority.

Hampton, Virginia - This plant was designed to process MSW at a design capacity, having disposed 210,000 tons during its first three years of operation.

Norfolk, Virginia - This facility processes approximately 140 TPD, operating seven days a week alternating between two boilers. Deterioration and rupture of several tubes in the side walls has been reported.

Portsmouth, Virginia - This plant continues operating, processing about 80 TPD.

The Dioxin Issue

Dioxin or TCDD is the generic name for a group of compounds known as polychlorinated dibenzo-p-dioxins. One form of dioxin, 2, 3, 7, 8 - TCDD, is considered among the most hazardous synthetic chemicals. In 1979, TCDD was detected in emissions from the Hempstead resource recovery plant in New York. Since then dioxin has been a serious issue for regulators and the public.

Presently, the Federal government does not regulate dioxin emissions from resource recovery plants. The authority to regulate such emissions clearly exists under the Clean Air Act. Emissions from municipal incinerators are indicated to be below the level of health risk. This position is however constantly being evaluated.

New York, for example, has developed a guideline standard for dioxin emissions from resource recovery facilities. The standard maximum annual ground level concentration of TCDD to 9.2 multiplied by ten to the negative power of 14 grams/m^3. This guideline value represents the highest ground level concentration of TCDD determined by EPA in its 1981 evaluation of health risks associated with dioxin emissions. The EPA reports those states in which the TCDD emissions from a sampling of five municipal waste combustors did not present a public health hazard for residents living in the immediate vicinity of the sites. The EPA health risk assessment also indicated that as long as emission levels of TCDD do not greatly exceed the emissions at the five US sites evaluated, there should be no reason for concern. Despite this EPA finding, there should be no reason for concern. Despite this EPA finding, project sponsors should anticipate utilization of the dioxin issue on the part of project opponents.

Human exposure to dioxin has so far been scientifically linked to only one health problem - a skin condition called chloracne, which tends to resolve quickly. Moreover, in situations where chloracne has developed, the exposure to dioxin was significantly greater than any exposure associated with emissions from resource recovery plants to date.

Perhaps the most extensive human experience with TCDD involves the town of Seveso, Italy. In 1976, TCDD was accidentally released from a chemical plant in Seveso. More than 37,000 persons are thought to have been exposed to varying degrees, making this the largest population exposure to TCDD. Other studies of persons exposed to dioxin have yielded similar results. But until these observations are confirmed by long term studies, no definitive conclusions can be drawn.

Dioxin while not harmless, is also not likely to be the threat that some might argue. Thus, while many questions about dioxin remain unanswered, a number of points can be stressed in addressing concerns about this chemical:

Dioxin emissions from resource recovery have not been shown to present a significant health risk.

Refuse incineration has been practiced worldwide for over a century without known adverse impacts on health.

New research indicates that better control over combustion conditions will allow some facilities to virtually eliminate dioxin emissions as well as other environmental risks.

CONCLUSION

The resource recovery industry operates in an ever changing environment. New standards of excellence are demanded in products and services as well as changing requirements. Identifying suitable recovery technologies is critical to the determination of recyclability for wastes that cannot be reused directly.

Incineration is the ideal alternative to landfilling as the final disposition for many solid wastes. Incineration has the advantages of 90% volume and weight reduction to an inert. Potential for recovery of heat energy and pollutants from the flue gas can meet environmental regulations. Identification and potential energy uses are primary concerns.

Waste to energy products in small and medium sized communities have a limited number of potential product markets. Thermal energy is typically a valuable and saleable product. It is advisable to identify and establish contact with industries with, ideally, the following characteristics:

A user of substantial quantities of process steam and/or hot water.

Internal generator of combustible wastes.

Thermal energy sales should be considered as an overall part of any resource recovery project.

The complexity of the environmental permitting process will vary based upon sized, design, location, state requirements and other factors. The nature of this process varies with applicants and the project.

Certain factors such as local opposition are beyond the control of the facility. Nevertheless, adequate planning, an open mind, and a co-operative attitude continue to be the most important factors for obtaining environmental permits and public acceptance.

Meanwhile the good news is that recycling and resource recovery has become an attractive waste disposal option, since it reduces waste handling costs, while conserving resources. The land disposal ban has changed the economics of waste management making recovery systems from waste disposal even more cost effective than before. **Tables 2** and **3** provide typical waste fuel parameter guidelines, and a comparison of municipal solid waste and coal fuel properties.

TABLE 2.

TYPICAL WASTE FUEL PARAMETER GUIDELINES

Waste Fuel Parameter	Preliminary
Ash	Not to exceed .2% by wt.
Total Halogen	Not to exceed 1,000 ppm (.1% by wt.)
Arsenic	Not to exceed 8. ppm.
Chromium	Not to exceed 3. ppm.
Mercury	Not to exceed 0.1 ppm.
Lead	Not to exceed 25. ppm.
Sulfur	Not to exceed .5% by wt.

TABLE 3.

COMPARISON OF MUNICIPAL SOLID WASTE AND COAL FUEL PROPERTIES

COMPOSITION	MSW	RDF	ILLINOIS COAL
C	25.6	33.8	57.6
H	3.4	4.5	3.7
O	20.3	26.2	5.8
N	0.5	0.6	0.9
S	0.15	0.2	4.0
ASH	24.9	11.9	16.0
MOISTURE	25.2	24.0	12.0
HEAT CONTENT BTU/POUND	4,450	5,750	10,000

CHAPTER 7

WASTE MINIMIZATION

INTRODUCTION AND OVERVIEW

The United States annually produces millions of tons of pollution and spends tens of billions of dollars per year controlling it. Significant opportunities to reduce or prevent pollution at the source through cost-effective changes in production, operations and use of raw materials exist for industry. Until recently industry and business was not overly concerned with costs resulting from waste generation. Source reduction opportunities were often not realized because of existing regulations, and resources required for compliance, focused on treatment and disposal, rather than source reduction. Regulations have heretofore not emphasized multimedia pollution management and have required technical assistance and information to overcome institutional barriers for source reduction adoption and implementation.

The Resource Conservation and Recovery Act (RCRA) as amended by the U.S. Congress, November 1984 states, "the Congress hereby declares it to be the national policy of the United States that, wherever feasible, the generation of hazardous waste is to be **reduced or eliminated** as expeditiously as possible. Waste nevertheless generated should be treated, stored or disposed of so as to minimize the present and future threat to human health and the environment."

Reduction of waste at the source has the potential for lessening the substantial amounts of the industrial waste generated every year. Examples of voluntary source reduction efforts made by specific companies show that major savings can be realized, both as a result of increases in production efficiency and in significantly reducing waste treatment, disposal, and clean-up. These incentives encourage industries to look for less expensive waste management alternatives. Little information or guidance is currently available to show them viable options. Federal assistance in the development and use of innovative source reduction techniques could help companies comply with pollution control laws while saving money and improving environmental protection.

Potential hazards posed by enormous amounts of toxic waste and the direct and indirect damage caused by its mishandling in the past have prompted extensive government and industry attempts at waste management. Congress has responded by passing increasingly stringent pollution control laws since 1970. At the same time, many industrial generators have voluntarily sought ways to improve their waste management practices. In theory, these efforts may employ options for handling waste: reduction at the source; recycling; treatment; and disposal.

In practice, waste management activities by both regulators and the regulated community have largely focused on treatment and disposal, and to a lesser extent on recycling. Although each of these techniques is appropriate in a comprehensive waste management strategy, government and industry are beginning to realize that end-of-the-pipe pollution controls alone are not enough. Significant amounts of waste containing toxic constituents continue to be released into the air, land, and water, despite stricter pollution controls and skyrocketing waste management.

Yet, there is increasing evidence of the economic and environmental benefits to be realized by reducing industrial waste at the source rather than managing such waste after it is produced. In testimony before the Subcommittee on Transportation and Hazardous Materials, Administrator Reilly of the EPA stated that "H.R. 1457, the Waste Reduction Act, is a big step in the right direction; the bill's emphasis on preventing pollution at the source rather than 'end of the pipe' control is appropriate and welcomed." Additional testimony received at Congressional hearings demonstrated that some manufacturing companies have modified their production processes, instituted better housekeeping procedures, and adopted other measures that have resulted in large percentages of waste being eliminated, while increasing production efficiency and lowering costs.

A key piece of legislation is the **Pollution Prevention Act of 1990**. The purpose of the bill is to make it the national policy to reduce or eliminate the generation of waste at the source whenever feasible and to direct the U.S.E.P.A. to undertake a multi-media program of information collection technology transfer, and financial assistance to the States to implement this policy and to promote the use of source reduction techniques. The bill is designed as a first step towards accomplishing pollution prevention techniques and encourages the voluntary reduction of hazardous wastes and other pollutants created during the manufacturing process.

Methods for Waste Reduction

Methods for reducing waste at the source are numerous, although they can be illustrated in several demonstrable categories. For example:

- **Manufacturing changes.** Atlantic Industries reduced its wastewater discharges by 55,000 pounds per year (while at the same time increasing product yield by 8 percent) by altering the chemical concentrations, lowering chemical reaction temperatures, and introducing a new method of combining dye components in the manufacturing process for its diazo dyes.

- **Equipment changes.** Dow Chemical significantly reduced its generation of hazardous chemical gases by substituting a pumping mechanism for the pressurized nitrogen gas it used to push raw materials from storage tanks into reactor vessels.

- **Product reformulations and substitutions.** Union Oil Company's chemical plant eliminated mercury wastes by substituting for a mercury biocide one that was mercury-free. Monsanto totally eliminated hazardous solid wastes from one step in its adhesives production by incorporing particulate matter-formerly filtered out as waste-in the final product.

- **Improved housekeeping.** An Exxon Chemical America facility cut the quantity of organic wastes entering its wastewater treatment plant by 75 percent simply by implementing a stewardship program, in which plant employees monitor discharges containing toxic constituents.

Potential for Source Reduction

The adoption of source reduction measures could prevent substantial amounts of environmental pollution. The Office of Technology Assessment (OTA), in its September 1986, report to Congress entitled "Serious Reduction of Hazardous Waste," estimated that hazardous waste reductions of 10 percent per year were achievable annually for the next five years. OTA concluded that there are substantial benefits to be realized by reducing the generation of hazardous waste through industrial process changes, raw material substitution, etc., and that a comprehensive, multi-media approach to reducing wastes going into the air, land, and water is needed.

In October 1986, EPA completed its "Report To Congress: Minimization of Hazardous Wastes." In that report, EPA concluded that mandatory standards of performance for waste minimization are neither feasible nor desirable but that EPA should establish a waste minimization program to develop information, to provide technical and information assistance to waste generators and States, and to consider long term options with respect to achieving waste minimization. EPA estimated that up to 33 percent of current hazardous waste generation could be eliminated.

In a follow-up report issued in June 1987, "From Pollution To Prevention," OTA compared the EPA report with its own September 1986 report on waste reduction. OTA concluded that both reports saw source reduction as:

- an option with many environmental and economic benefits compared to management and regulatory options that deal with waste that is already generated;

- technically and economically feasible with current science and technology; and

- is in only limited use by industry because of a number of institutional obstacles in both industry and government.

Barriers To Source Reduction

Increasing concerns over materials costs, pollution control laws and regulations, potential company liability, and public relations, have helped to create a situation where substantial levels of voluntary source reduction may be achieved while meeting both the environmental and the economic interests of industrial generators. Yet, according to OTA, for the majority of companies little source reduction is being achieved.

Predominant factors inhibiting the increased use of source reduction techniques as a strategy to reduce pollution have little to do with technological constraints, economic costs, or governmental prohibitions. In fact, lack of awareness or lack of information has been cited as the most common reason more companies do not practice source reduction. Many company managers simply are not aware of techniques that could be applied to their business operations. Business owners and regulators are frequently so oriented to meeting the requirements and deadlines of current pollution control regulations and statutes that they do not seek out innovative alternatives.

In its 1986 report, EPA cited an array of disincentives to source reduction. Aside from the feasibility aspects, awareness of upper-level management was one of the most influential factors in the introduction of source reduction programs. EPA states that "to be successful in bringing about changes in plant operation and/or design, policy-making and implementation processes within, companies are largely dependent upon upper management support."

EPA Waste Reduction Programs

EPA has established only a limited waste minimization program for hazardous wastes regulated under RCRA. Under Section 3002 of RCRA generators of hazardous waste are required to report to EPA on a biennial basis the amount of hazardous waste they generate, the efforts they have undertaken to reduce such waste, and the amount of the reduction. RCRA hazardous waste generators are further required to certify that they have in place a program to minimize the amount of hazardous waste they generate.

EPA has not established comparable source reduction programs for the air emissions regulated under the Clean Air Act or the wastewater discharges regulated under the Clean Water Act. For this reason EPA is currently unable to determine accurately whether similar source reduction efforts are being made with respect to air emissions or water discharges, or whether RCRA source

reduction gains are being made at the expense of air emissions or water discharges.

Under Section 313 of the Emergency Planning and Community Right to Know Act of 1986, EPA began obtaining data on the release of certain toxic substances into the environment. Industries that manufacture, process, or use certain toxic chemicals above the specified threshold amounts are required to report all releases of such chemicals into the air, water, and land. In these Sections 313 reports, EPA has provided for optional reporting of any source reduction and recycling efforts undertaken by industry with respect to such toxic chemical releases. Under current law, however, industries are not required to provide this information. EPA therefore does not know what level of reporting with respect to waste reduction to expect, although the agency estimates that relatively few companies will voluntarily attempt to supply the information.

State Programs

A number of States have already instituted source reduction programs. Most of these State programs are designed to provide "hands-on" technical assistance to industry in applying source reduction techniques. A few of the State programs provide financial assistance to industries interested in source reduction and provide information designed to promote the benefits of source reduction. Some of the State programs also include waste-end taxes, loan and grant programs, information exchanges, and tax bonuses as additional disincentives or incentives to source reduction.

According to OTA, many State waste reduction programs may be responding only to problems of land disposal of RCRA-defined wastes, failing to consider that some efforts merely succeed in transferring the pollution from land to air or water. Generally, State waste reduction programs lack the funding and specific commitment to source reduction to achieve their full potential.

In its 1986 report, OTA found that "current State programs are not substantially increasing waste reduction nationwide." In fact, most State programs are directed only at small business, which may account for only a small fraction of the waste produced within the State. OTA concluded that "although States have led the Federal Government in actively promoting waste reduction, a parallel Federal effort is needed to raise waste reduction to a stature comparable of that of pollution control."

WASTE MINIMIZATION DEFINITIONS

Waste Minimization - Waste minimization can be defined as the means by which we achieve a reduction of hazardous waste that is generated or substantially treated, stored or disposed of. It includes any source reduction or recycling activity undertaken by a generator that results in either

- the reduction of total volume or quantity of hazardous waste or

- the reduction of toxicity of the hazardous waste, or both, so long as such reduction is consistent with the goal of minimizing present and future threats to human health and the environment (EPA's Report to Congress, 1986, EPA/530-SW-86-033).

Source Reduction - This refers to any activity that reduces or eliminates the gene-

ration of hazardous waste at the source, usually within a process (op. cit.).

Recycling - A material is "recycled" if it is used, reused, or reclaimed (40 CFR 261.1(c)(7)). A material is "used or reused" if it is either

- employed as an ingredient (including its use as an intermediate) to make a product; however a material will not satisfy this condition if distinct components of the material are recovered as separate end products (as when metals are recovered from metal containing secondary materials) or

- employed in a particular function as an effective substitute for a commercial product (40 CFR 261.1(c)(5)). A material is "reclaimed" if it is processed to recover a useful product or if it is regenerated. Examples include the recovery of lead from spent batteries and the regeneration of spent solvents (40 CFR 261.1(c)(4)).

COLLECTING AND COMPILING DATA

The questions that information gathering efforts attempt to answer include the following:

- What are the waste streams generated from the plant? And how much?

- Which processes or operations do these waste streams come from?

- Which wastes are classified as hazardous and which are not? What makes them hazardous?

- What are the input materials used that generate the waste streams of a particular process or plant area?

- How much of a particular input material enters each waste stream?

- How much of a particular input material enters each waste stream?

- How much of a raw material can be accounted for through fugitive losses?

- How efficient is the process?

- Are unnecessary wastes generated by mixing otherwise recyclable hazardous wastes with other process wastes?

- What types of housekeeping practices are used to limit the quantity of wastes generated?

- What types of process controls are used to improve process efficiency?

This information can be useful in conducting the assessment. Reviewing such information will provide important background for understanding the plant's production and maintenance processes and will allow priorities to be determined.

Waste Stream Records

One of the first jobs of a waste minimization assessment is to identify and characterize the facility waste streams. Information about waste streams can come from a variety of sources. Information on waste quantities is readily available from completed hazardous waste manifests, which include the description and quantity of hazardous waste

shipped. The total amount of hazardous waste shipped during a one-year period, for example, is a convenient means of measuring waste generation and waste reduction efforts. However, manifests may lack such information as chemical analysis of the waste, specific source of the waste, and the time period during which the waste was generated. Also, manifests do not cover wastewater effluents, air emissions, or nonhazardous solid wastes.

Other sources of information on waste streams include biennial reports and NPDES monitoring reports. These NPDES monitoring reports will include the volume and constituents of wastewaters discharged. Toxic substance release inventories prepared under the "Right-To-Know" provisions of SARA Title III, Section 313 (Superfund Amendment and Reauthorization Act) may provide valuable information on emissions into all environmental media (land, water, and air).

Analytical test data available from previous waste evaluations and routine sampling programs can be helpful if the focus of the assessment is a particular chemical within a waste stream.

Information required for waste minimization assessments include:

Design Information

- Process flow diagrams

- Material and heat balances (both design balances and actual balances) for production processes and pollution control processes

- Operating manuals and process descriptions

- Equipment lists
- Equipment specifications and data sheets
- Piping and instrument diagrams
- Plot and elevation plans
- Equipment layouts and work flow diagrams

Environmental Information

- Hazardous waste manifests
- Emission inventories
- Biennial hazardous waste reports
- Waste analysis
- Environmental audit reports
- Permits and/or permit applications

Raw Material/Production Information

- Product composition and batch sheets
- Material application diagrams
- Material safety data sheets
- Product and raw material inventory records
- Operator data logs
- Operating procedures
- Production schedules

Economic Information

- Waste treatment and disposal costs
- Product, utility, and raw material costs
- Operating and maintenance costs
- Department cost accounting reports

Other information

- Company environmental policy statements
- Standard procedures
- Organization charts

Flow diagrams provide the means for identifying and organizing information that is useful for the assessment. Flow diagrams are prepared to identify important process steps and to identify sources where wastes are generated. Flow diagrams are the foundation upon which material balances are built. Material balances are important for many waste minimization projects, since they allow for quantifying losses or emissions that were previously unaccounted for. Also, material balances assist in developing the information.

THE SYSTEMATIC APPROACH

Waste reduction is a continuing effort. It is necessary to trace back to each source for characterization, instead of dealing with a mixed flow. Through process observation and sampling, determine the quantity, composition, and flow at all times of the day and night. The **variation of waste flow and concentration is very important**. This process observation and sampling will provide insight into the true cause of waste generation for later selection of waste reduction alternatives. It provides information on whether wastes are generated due to process chemistry, equipment used, operating procedure, or poor housekeeping. Cost figures using a multimedia approach in cost estimation will reflect the true cost to achieve the kind of treatment that will not result in moving the pollution problem from one media to another.

Sources can be grouped according to physical and chemical characteristics, recoverability, and potential market value. Following the evaluation and selection of alternatives, a proposal may be made for plant test or lab/pilot study, depending on the readiness of the technology involved. If lab/pilot studies are needed, results of the study are plugged back into the evaluation stage. Coming out of this loop will be the best solution for the sources for project implementation.

Such a step-by-step waste reduction approach is primarily aimed at existing processes. When dealing with new processes, the waste reduction concept should be applied starting from early stages of process development. Emphasis in raw material substitution, process modification and recycle/reuse of waste streams should be made from the beginning. Environmental costs should be added to the traditional technical and production cost considerations for comparison of alternatives at different stages of process development.

OPTIONS AVAILABLE

Housekeeping - Although this is obvious, improved housekeeping is often the most effective first step for smaller companies.

Material Substitution - This solution, which involves replacing the pollution forming ingredient in a product with a less toxic one, may be the most logical and complete way to eliminate waste but may be applicable only in certain situations.

Recycling and Reuse - Many industries have been able to invest in on-site recycling equipment, and thus solve or reduce both management cost problems and raw material needs. For the small industries that can't afford such investments, there are numerous off-site recyclers who will take the wastes. Recovery processes include:

Distillation - mainly used to recover organic solvents. A unit distills the solvent and recondenses the solvent vapor for reuse. A typical distillation unit may recover 85 to 90 percent of the original solvent.

Absorption - a bed of carbon or other compound is used to absorb gaseous materials or evaporating solvent vapors. The compound can be recovered in a highly concentrated form when the absorbent beds are regenerated, usually with the application of steam. Absorption units can achieve 90 percent or better solvent recovery.

Filtration - waste streams containing particulate matter can be passed through filters, which capture the particles for reuse. The porosity of the filter will determine which particles are recovered. Some industries add lime to their waste stream before passing it through the filter; in this scheme heavy metals are precipitated and more easily recovered. Reverse osmosis as well as other techniques are proving useful.

Electrolysis - an electrically charged cell attracts ions and removes them from the waste stream. The electroplated metal can then be removed from the cathode and sold as scrap. Many metals can be removed in this way, including copper, manganese, nickel, chromium, and silver. The recovery rate is excellent, ranging from 90 to 99.9 percent.

Waste Exchange - materials that one industry views as waste may actually be a valuable resource to another. Informational networks called "waste exchanges" operate like the "personals" of popular newspapers: industrial firms fill out forms describing the waste they produce or the raw materials they need. This information is then printed in publications. The search is complete when a firm is contacted by someone who needs and is able to recycle the raw materials present in the waste. Both parties benefit, either from the sale and in avoiding disposal costs of previously unmarketable wastes, or from receiving precious raw materials at inexpensive prices.

Detoxification - such processes neutralize hazardous waste into harmless materials. However, these are the least desirable of solutions because the waste still remains after the detoxification process has been completed. Detoxification should be considered only after the previous approaches have been tried. Detoxification includes any of the following approaches:

- Chemical neutralization - an acid waste, for example, can be used to neutralize a basic waste, thereby rendering them both saleable. Or a chemical can be used to break down a hazardous compound, as is done with cyanide in electroplating waste solutions.

- Biological - treatment with microorganisms can break down many organic compounds in some hazardous waste.

- Thermal treatment - various thermal detoxification methods, including incineration and wet oxidation, are now available for detoxifying organic compounds. Incineration is useful because, in addition to destroying the waste, it also generates heat energy that can be put to a number of uses. One must insure that toxic substances will be completely destroyed in the incineration process. If the incinerator is not capable of functioning properly, the company may be liable.

ASSESSMENT OVERVIEW

In order to develop a comprehensive list of waste minimization options for a facility, it is necessary to understand the sources, causes, and controlling factors that influence waste generation. Typical causes and sources of waste include:

- **Material Receiving**

  Loading docks, incoming pipelines, receiving areas.

  Packaging materials, off-spec materials damaged containers, inadvertent spills, transfer hose emptying.

- **Raw Material and Product Storage**

  Tanks, warehouses, drum storage yards, bins, storerooms.

Tank bottoms; off-spec and excess materials; spill residues; leaking pumps, valves, tanks, and pipes; damaged containers, empty containers.

- **Production**

Melting, curing, baking, distilling, washing, coating, formulating, reaction.

Washwater; rinse water; solvents; still bottoms; off-spec products; catalysts; empty containers; sweepings; ductwork clean-out; additives; oil; filters; spill residue; excess materials; process solution dumps; leaking pipes; valves, hoses, tanks, and process equipment.

- **Support Services**

Laboratories - Reagents, off-spec chemicals, samples, empty sample and chemical containers.

Maintenance shops - Solvents, cleaning agents, degreasing sludges, sand-blasting waste, caustic, scrap metal, oils, greases.

Garages - Oils, filters, solvents, acids, caustics, cleaning bath sludges, batteries.

Powerhouses/boilers - Flyash, slag, tube clean-out material, chemical additives, oil empty containers, boiler blowdown, watertreating chemical wastes.

Cooling towers - Chemical additives, empty containers, cooling tower bottom sediment, cooling tower blowdown, fan lube oils.

PRIORITIES

No matter how successful a waste reduction program is, there will always be some wastes to be treated or disposed of. For particular streams, the following priority (in the decreasing order of preference) should be applied in pollution control:

- Reduction at the process
- Material recovery and reuse
- External utilization
- Waste treatment
- Waste/residue disposal

The last two items fit into the category of "END-OF-PIPE" approach. The first item can be achieved by good housekeeping and waste accountability, improved operating procedures/conditions and process modification or equipment design.

CHAPTER 8

WORKING WITH HAZARDOUS MATERIALS

INTRODUCTION AND OVERVIEW

Any health and safety program in a work environment should define the exposure of workers to chemical, biological hazards and exposure to physical stresses (e.g., noise, heat, mechanical dangers) as well as affording the necessary safeguards and protection. Evaluating the occupational environment requires recognizing potential hazards, conducting field surveys and data interpretation prior to implementing protection and control methods.

Work environments have either potential or actual hazards requiring recognition, identification, measurement, monitoring and/or protection. Risks or hazards require knowledge of raw materials; how materials are handled and modified during processing; materials evaluation; and conditions that exist during manufacturing.

The first step in recognizing environmental factors from which occupational hazards arise is familiarization with the materials, plant and work operations. Processing and handling must be assessed under normal conditions as well as anticipated or even catastrophic conditions. The potentially hazardous nature of conditions and materials must be determined before, during and after their use. The maintenance of a safe, healthy work environment depends upon thorough familiarity with materials and conditions in the workplace. Working with hazardous materials and wastes is serious business and handling and exposure to such materials pose significant threats to workers and the environment.

Hazardous materials and wastes are a serious safety and health problem that continues to endanger human and animal life and environmental quality. Hazardous wastes are discarded chemicals that are toxic, flammable or corrosive, and they can cause fires, explosions, and pollution of air, water and land. Unless hazardous materials and waste are properly handled, treated, stored or disposed of, it will continue to do great harm to all living things that come into contact with them now or in the future.

MATERIALS

Knowledge regarding **raw materials** handled and used, and the nature of products and intermediates are essential to any industrial hygiene or safety program. Assessing potential material problems should also include information on the impurities present as well (for example, some solvents may contain trace contaminants that are potentially toxic). These should be taken into account in evaluating possible risks.

By-products, intermediates, and **final product** form from raw materials and offer potential hazards. Undesirable by-products such as carbon monoxide from combustion is an example. Lead oxide fumes from lead paint-coated steel during welding or torch cutting operations is another example with vaporized lead combining with oxygen during the high temperature exposure. Chemical breakdown of chemicals can occur to produce toxic fumes or irritants as in the case of hydrogen chloride generation from the degradation of chlorinated hydrocarbon solvents such as trichloroethylene or perchloroethylene.

A first step in any assessment is to list all materials in use at the plant. The first source of information on these materials should be their supplier. Vendors should provide data which includes: properties; composition; recommended procedures for storage and handling; and special measures such as safety gear and ventilation, if required. The **toxicity** of a substance may not be the only standard or necessarily the most important one for consideration. Possibilities of reactions (physical as well as chemical) with other materials may exist. Concentration build-up potentials may be present which will require ventilation control or enclosures.

Too much information is never enough. Compositional analysis of chemical constituents in materials being handled is a paramount requirement. This is needed to assess the hazard. Use of pure or simple chemicals are more easily identified than mixtures of proprietary products or trade names. Identification in the latter case may become more difficult and therefore using information between suppliers should be used guardedly. Recognition of the basic hazard, whether it is physical/chemical/or biological,

is the first step in any assessment and there is no substitute for knowledge and accurate data.

Initial assessment based on materials' properties and processes will determine the course of sampling, measurement, and **monitoring programs.** The importance of these programs is basically:

- Analytical data may be required for designing control equipment.

- Instrumentation may be required for process control.

- Data may be required for submission to government enforcement agencies for compliance.

- As legal evidence of compliance or in a court of law.

- For protection of health and property for both workers and the public at large.

In order to assess the impact any material may have on a work area environment or human health, it is necessary to have data available in a number of areas. These include:

- physical/chemical properties;
- degradability/reactivity;
- accumulation potential;
- uses/likely exposure routes and levels;
- amounts produced; and
- toxicity.

A systematic approach should be followed to identify environmental risks with the types of products and materials existing in the process. Allowable levels and toxicological effects may be determined based on exposure guidelines from OSHA. An understanding of the process and its variations are the factors to consider next.

Toxicity and hazard severity depend upon both the level or concentration of a material and the duration of exposure. This requires, certainly as a second step, a process analysis and detailed examination of the flow sheet or operation. Information should include potentially abnormal as well as usual or standard operating conditions and all factors that might affect worker exposure.

As well as the materials involved in any process, the type of operation will also determine if a potential hazard exists and typically the controls that may be required.

A **process flow sheet analysis** should be constructed. This requires:

1. A sequential showing of how each material is introduced into the process.
2. Products and by-products from each step.
3. The flow of materials and their quantities.
4. An accurate material balance.
5. Instrumentation and equipment controls with safety controls or features indicated as to location and operating ranges.

This should be accompanied by standard operating procedures and instructions and descriptions of the operations involved. In many industrial operations several different hazards may exist simultaneously, so that it is necessary to study the complete process to assess all of the potential risks.

Manufacturing may involve both physical and chemical hazards. The **physical hazards** associated with operating conditions include high temperature and pressure, mechanical, and electrical safety. **Chemical hazards** are those associated with processes which may contain flammable, toxic, and/or highly reactive materials. Chemical reactions are not usually first order, with normally a number of competing side reactions in addition to the desired reaction and the formation of intermediate compounds before resulting in the final product. It is therefore important to know what these compounds are and evaluate the associated hazards. The analysis approach should develop information from a consideration of reaction chemistry and a knowledge of feedstock impurities.

Mechanical Hazards - It is important to appreciate that in addition to hazards that may result from environmental factors caused by materials, processes and operations, safety and health may be endangered by mechanical hazards.

Work Operations Survey - A survey of the manufacturing operations in the form of a physical **walkthrough inspection** is an early requirement in existing facilities. This may be done on an initial basis and is an ongoing requirement of periodic inspections to determine if items such as required monitoring, housekeeping, and maintenance are in effect. First, surveys and studies as well as continuing inspections will establish effectiveness of control measures, investigate complaints, and generally determine compliance with safety policy, programs, and regulations.

Preliminary surveys are used to establish a data base to which future conditions can be compared against or evaluated. Before such a survey, an adequate review of materials and processes should have been made as previously described which includes raw materials, by-products, products, and equipment as well as process operations and conditions such as temperatures, pressures, etc.

Any **physical plant survey** should include an inspection of all operations while in progress. Operations which are not continuous or are run only once in a while may represent some large risks or high degree of hazards. Night shift or weekend work must be included to see if there are any variations at these times since production rates or cycle times during the process may be significantly different.

Observations should yield the following types of data:

- number of employees possibly exposed
- exposure time
- materials used and handled
- handling procedures
- floor or process area storage of materials/products
- housekeeping
- atmospheric conditions in the work area
- availability of ventilation
- protective equipment utilized by workers
- possible contaminant contact to the worker (possible route of entry to the body).

Plant level functions and **data collection** functions for ongoing operations should give the following outputs:

- Occupational injury accidents
- Property damage accidents

- Environmental incidents
- Machine and process statistics
- Time/date of each failure
- Failure description
- Time to correct each failure
- Time/date of operation turn-on and turn-off
- Delayed events associated with accidents or incidents that occurred in the past

SAFETY ITEMS FOR WHICH VIOLATIONS SHOULD BE AVOIDED

- Limited access to portable fire extinguishers, fire hoses, and sprinkler heads blocked

- Limited emergency egress - inadequate aisleways, stairways, fire escapes, and exits

- Excessive floor loadings

- Excessive noise/vibration

- Machinery without adequate guards

- Electrical tools, equipment receptacles without proper electrical grounds

- Lack of safety valves or blocked valves on high pressure stream lines

- Unprotected floor or wall openings

- Improper storage of flammable materials

- Lack of readily accessible emergency showers near hazardous/corrosive chemicals

- Lack of canopies on fork-lift trucks

- Inadequate fire protection systems

- Exposed energized electrical equipment

- Lack of identification on pipes, electrical switches, breakers

- Inaccessible shut-off valves

- Unmarked traffic lanes for inside vehicle traffic

- Lack of proper guardrails

- Lack of 3-wire electrical outlets

- Ladders not equipped with safety shoes

- Inadequate lighting for work and movement

- Overhead obstructions offering safety hazards

ABNORMAL OPERATIONS

Basic approaches in identifying potentials for abnormal operations or conditions that might be conducive in leading to high-risk high-hazard situations can include several approaches. One is a retrospective approach, the other a prospective approach. The retrospective analysis focuses on historic losses and incidents and study past conditions/losses, through study to identify key factors to be changed to avoid future hazards or risks. The prospective approach predicts conditions from situations for which no past data are available.

Where **past data** is available, retrospective analysis of situation and accident details can be used to identify severity factors that led to the hazard situation. Once the factors are identified, corrective action can be taken. For example, analysis of **compensation claims** of employees might show occurrence with persons who had similar histories of prior injuries. This could lead to identification of specific prior injuries or medical conditions that could be used as specific prior injuries or medical conditions that could be used as a specific severity factor. Good safety practice would then suggest a program developed to identify employees with such records and assign them to work duties where the risk factor would be zero.

Where no data are available on prior accidents, the prospective approach can be used to identify areas of potential hazards and risk for corrective action. Surveys of existing or planned facilities are made to identify **severity indicators**, such as presence of hazardous material or conditions. Here again, where one or more situations are indicated, good safety practice will determine methods to reduce or eliminate potential risks. In many situations, severe indications can be readily identified due to the presence of large amounts of hazardous materials for example while other risks may be more subtle and long range.

Many models and analytical techniques have been developed to assist in performing this task. The more widely used techniques include: Fault Tree Analysis and Technique for Human Error Prediction; Failure Mode and Effect Analysis. These usually require extensive, time-consuming analysis for even the simplest projects and offer difficulty in handling some of the logical interactions which can occur among events. However, there may be certain types of projects for which such models can be developed.

Problems related to **risk-hazard safety analysis** typically encountered are:

- limited availability of data
- problems of small numbers
- reliability of collected data

Quite often limited safety staffs in the field are hard pressed to keep up with safety problems, leaving limited or little time for data collection and reporting. Overcollecting and underutilizing data are commonplace. Large amounts of environmental monitoring data may be collected with only small portions in usable form.

Examples of potential safety risks include:

- high pressure steam or air tanks
- concentrations of flammable liquids
- run-away/out-of-control chemical reactions
- intense ionizing radiation sources
- experimental biological substances
- mutagenic or teratogenic materials
- large, nearby work population
- high-value or difficult-to-replace equipment
- heavy dependence by process on a specific person or piece of equipment or a facility
- lack of sprinklers, alarms
- lack of protective clothing
- lack of engineered safety features (e.g., interlocks, instrumentation, safety valves, rupture discs)
- inadequate or non-existent barriers to release of hazardous material
- improperly designed barriers which may themselves become hazards if an accident occurs (e.g., brick or block wall used to protect workers from an explosive hazard. Wall parts could become missiles.)
- non-routine operations

- mismatch of skills with job complexity
- lack of after-care procedures for accidents
- lack of organized and trained medical and fire units
- lack of defined responsibility for operations before, during, or after accidents
- inadequate communications.

TOXICITY

The term **toxicity** is a relative term and for a material/substance/chemical depends on composition and basic properties as well as dose, route of body entry, exposure conditions, susceptibility of the individual exposed. To define a material as **toxic** or **nontoxic** depends on use and exposure.

The National Institute for Occupational Safety and Health (**NIOSH**) has defined a **toxic substance** (Christensen, H.E., ed. "Registry of Toxic Chemical Substances"-1975 ed. (formerly "The Toxic Substance List"), NIOSH, PHS, DHEW, June 1976) as:

A toxic substance is one that demonstrates the potential to induce cancer, tumors, or neoplastic effects in man or experimental animals; to induce a permanent transmissible change in the characteristic of an offspring from those of its human or experimental animal parents; to cause the production of physical defects in the developing human or experimental animal embryo; to produce death in animals exposed via the respiratory tract, skin, eye, mouth, or other routes in experimental or domestic animals; to produce irritation or sensitization of the skin, eyes, or respiratory passages; to diminish mental alertness and reduce motivation, or alter behavior in humans; to adversely affect the health of a normal or disabled person of any age or of either sex by producing reversible or irreversible bodily injury or by endangering life or causing death from exposure via the respiratory tract, skin, eye, mouth, or any other route in any quantity, concentration, or dose reported for any length of time.

CHEMICAL INFORMATION REQUIREMENTS FOR ENVIRONMENTAL, SAFETY AND HEALTH HAZARD ANALYSIS

Substance Identification

Descriptive Identification

- Nomenclature (CAS Registry Number, CAS Preferred Name, Synonyms, Trade Names, Other Codes)

- Chemical Structure (chemical structure/molecular formula/formula weight)

- Composition (determination methods/impurities)

Chemical Properties

- Corrosivity in terms of pH
- Reactivity (with: water/oxidation-reduction/acid/base/photoreactivity/nucleophilicity/electrophilicity/thermal)

Physical Properties (state/color/texture/density/index of refraction/melting point/boiling point/freezing point/flash point)

- Volatility (vapor pressure/vapor density)

- Solubility (water/organic solvents/octanol-water partition coefficient)
- Special Properties (absorption spectroscopy/NMR spectroscopy/flourescene spectroscopy/optical rotation/X-ray diffraction/mass-spectroscopy)
- Persistence (half-life)
- Methods of Identification (analytical techniques/standard protocols)

Production Aspects

- Production Source (control technology/by-products)
- Handling (storage/transport/fire rating)

Occupational Exposure

- Total Work Force/Occupational Group
- Duration and Frequency
- Route of Exposure

Biological Effects

- Clinical Studies
  - Exposed populations
  - Procedures
  - Results
- Toxicology (Human/Animal)
  - Acute toxicity (study characteristics)
  - Sub-chronic toxicity
  - Chronic toxicity (carcinogenicity/teratogenicity/mutagenicity/other)
- Metabolism (Human/Animal)
  - Blood and other organ levels
  - Excretion rates
  - Absorption (skin/respiratory tract/gut)
  - Distribution (organs/tissue sites)
  - Chemical interactions

Standards and Regulations

- Federal Standards and Regulations (CFR)
- State Standards and Regulations
- Local Standards and Regulations
- Non-U.S. Standards and Regulations
- International Standards and Regulations

HAZARD SUMMARY

Potentially **hazardous situations** may be a result of the following type incidents:

- Fire
- Explosion
- Toxic vapor, fluid, or dust release
- Corrosive vapor or fluid release
- Overexposure to noise
- Electrical shock
- Release of cryogenic fluids
- Falls of personnel
- Fragmentation of a rapidly spinning grinding wheel, etc.
- Personnel eye exposure to arc welding operations
- Sudden release of pressurized fluids (gases)
- Overexertion
- Exposure to a mechanical power transmission apparatus
- Exposure to other mechanical equipment
- Violent chemical reaction

HAZARD EVALUATION

Toxicity of a material is not synonymous with the **hazard** represented. Toxicity is the inherent capacity of a substance to produce injury and is a function of the dose level or the duration of the exposure. Hazard is the possibility that a substance will produce an adverse effect when an amount of material is encountered under specific conditions. Hazard is therefore the combination of both the inherent toxicities of a material and the risk of exposure.

In the evaluation of a workplace health hazard, the factors to be considered include:

- Quantity of material in contact or in the metabolism system required to produce injury.
- Probability that the material will produce injury.
- Rate at which the material is generated.
- Control measures limiting the exposure.

Hazard and toxicity **evaluation** have many aspects. **Exposure** to materials and situations on the work area as well as the environment in general can be divided into a number of exposure classifications which are:

Acute exposures - where the dose is a single event and rapid. While this type exposure is usually short the results or effects may be irreversible. Acute exposure is a single dose in which protective mechanisms of the body are overcome.

Subacute exposures closely resemble acute exposures except they result from 12 to 40 doses (or exposure) and may be frequent, repeated, or extended over hours or days.

Chronic exposures - involve exposure or dose delivery over a period of time. Time and frequency concern of such exposures are a function of a particular material's toxicity, dose, and effects. Such exposures can occur at varying frequencies and can become continuous. Chronic exposures are characterized by long duration low level, where the body capacity for detoxification is exceeded over a period of time.

Long-term exposures - are typically exposures of more than one year on a continuing basis. This type of exposure is rare in the work environment and may usually occur on environments outside the plant or is employed in laboratory techniques for animal testing.

Toxicological interests in health safety fall into three main categories: environmental; economic; and forensic. Environmental concerns pollution, residues, and industrial hygiene. Economic areas involve drugs, food additives, pesticides, and insecticides. Forensic areas include diagnosis, therapy and medical/legal aspects.

EXPOSURE

Exposure is the length of time an individual is in contact with a substance or situation. As we have seen, the period of time may be short, usually referred to as being acute, or long, referred to as being chronic. The next most important questions are how an individual comes in contact with the material or what is the **route of entry** such as ingestion through the gastrointestinal tract, skin absorption or inhaled by breathing? Of the three entry methods, **inhalation** is regarded as the most important because it is the most frequent entry route. Substances suspended in air in the form of fumes, mists, vapors can enter through the nose and mouth to be deposited on the lungs. Some of these materials pass through the alveoli directly into the blood, others may be acted upon by the gastrointestinal tract or lymphatic system.

As in a chemical reaction, the rate at which materials enter the body depends on concen-tration, and the dose to which an individual is exposed and how fast the material can pass through the body's cell membranes and protective mechanisms. Generally, the most commonly affected organs by chemicals in the blood are the **liver** and **kidneys**. These organs have an ability to extract many chemicals and the liver has a high capacity to metabolize them. Two other areas of concern are **bone** and **fat cells**. Chemicals such as **DDT** have been found in fatty tissues and more than 85 percent of the lead in the body is found in the bone structure (skeleton).

The possibility of **special susceptibility** among certain workers is a question almost impossible to answer using occupational exposure data. Unless sufficiently large numbers of workers are affected, there is a chance that the most susceptible groups may not have been exposed. Increased susceptibility among groups includes those described above and may represent only a limited or small potential risk when considered in a large population. Nevertheless, the risk is present and should be evaluated and reduced.

A **no effects level** is the level below which no undesirable results occur on human health and the normal protective mechanisms of the body are not overloaded. In the strictest sense a no effects level does not exist since there are always slight deviations within normal biological change which are not necessarily harmful to health. The problem therefore of defining allowable deviations and identifying hazards and risks in the work-area environment is complicated to say the least. For example, is a **physical irritation** such as itching of the skin or eye tearing, a nuisance or in combination with other environmental agents a predisposition to further response and damage? Should the

irritation be considered an adverse effect although its presence may not be a health impairment?

Acute or chronic **illness** may be caused by inhalation, absorption, ingestion, or direct contact and can result in damage in the categories listed below:

- **Dust disease** of the lung (**pneumoconioses**) - examples include **silicosis**, **asbestosis**, coal worker's pneumoconioses, **byssinosis**, and other pneumoconioses.

- **Respiratory conditions** due to toxic agents - examples include pneumonitis, pharyngitis, rhinitis, or acute congestion due to chemicals, dusts, gases or fumes.

- **Poisoning** (**systematic effects** of toxic materials) - examples include **poisoning** by lead, mercury, cadmium, arsenic, or other metals; poisoning by **carbon monoxide, hydrogen sulfide, benzene, carbon tetrachloride**, or other **organic solvents**; poisoning by **insecticide sprays** and other chemicals such as **formaldehyde, plastics, resins**, etc.

- Occupational **skin disease** or disorders - examples include contact **dermatitis, eczema** or rash caused by primary irritants and sensitizes; chemical burns, or inflammations.

- Disorders due to repeated **trauma** - examples include **noise** induced **hearing loss**; synovitis and bursitis, Raynaud's phenomena; conditions due to repeated motion/vibration/pressure.

Responses to toxic substances can vary considerably and between individuals. Problems with toxic substances that cause dysfunction of an organ are referred to as systemic poisoning. Often there is a lag time between exposure and onset of disease or illness. Exposures to air with concentrations of **oxygen less than normal** (16-19.5 percent by volume) can produce death by asphyxia).

Typical Responses, Examples and Results to Toxic Exposure

- Behavioral - e.g., Oxygen deficiencies/loss of perception

  Mercury/emotional problems

- Biochemical - Organophosphorus insecticides/prohibit acetylcholinesterase/result in nervous tremors

- Carcinogenic - Vinyl chloride monomer/cancer of the liver

- Mutagenic - DNA and gene changes in infants

- Pathological - Silica/lung tissue deformed

- Physiological - Ammonia gas/decreased pulmonary function

- Reproductive - Cadmium/affects reproductive organs and their ability to function properly

- Teratogenic -
  Thalidomide/damage to a developing offspring in a pregnant female

Another problem in assessing occupational **exposure data** is the irregularities of exposure in the workplace. Concentrations of chemical agents in air, for example, may be inconsistent. Fluctuations in duration of exposure as well as concentration occur more often than not. Also, workroom exposures are seldom pure exposures. Mixed exposures of different chemical agents in the workroom environment may make it difficult to get precise values for any one substance. The existences of mixtures present the possibilities of interactions as well as synergistic health effects.

PROCESS AND MATERIAL HAZARDS

Hazards can be characterized into chemical hazards and equipment hazards. Chemical hazards are related to the physical and chemical properties of materials present in the process. Important properties with regard to hazard potentials of a chemical substance are its flammability, explosiveness, toxicity, and corrosivity. Equipment hazards are related to materials of construction, equipment age, maintenance, instrumentation and controls, and human error. Equipment hazards are site specific and must be referenced to a specific plant.

Material hazards include:

- oxidizing materials

- materials that react with water to produce combustible gases

- materials subject to spontaneous heating

- materials subject to spontaneous polymerization

- materials subject to explosive decomposition

- materials subject to detonation.

Process hazards include:

- loading and unloading operations

- handling flammable materials in open systems

- continuous reactions

- batch reactions

- potential for contamination

- low-pressure operations (less than atmospheric pressure)

- operation in or near explosive range

- low temperature in carbon steel vessels

- operation above flash point

- operation above autoignition temperature

- high pressure operations (greater than 17 atmospheres)

- difficult to control reactions

- explosive hazard

- large quantity of flammable liquid.

SAMPLING AND MONITORING

Any program studying the occupational environment invariably must look at the **workroom atmosphere**. This is particularly true if operations involve potential **air contaminants** as a result of storage, handling, processing, or packaging materials. Factors in conducting a field study or monitoring program all relate to sampling:

- Where (point/room location/breathing zone)

- When (time of day/shift)

- How long (minutes to days)

- How many samples (grab or continuous)

Once the survey is conducted, results have to be interpreted and compared to health standards, previous data, and recommendations made if necessary for corrective measures. Sampling instruments can fall into one of three general categories: direct reading; containment removal from a measurable air quantity; collection of a known volume for laboratory analysis off-site. Since no single or universal sampling instrument is available the choice will depend on type of analysis/information required; ease of use; reliability; efficiency; sensitivity/accuracy/reproducibility.

Occupational environment assessment strategies at various stages of study have already been outlined for identifying process, stream, area, and operational priorities. Sampling and analysis techniques will establish actual conditions and can be done on a staged or priority selection basis depending on the type of format and data required. Output from sampling and analysis is re-

quired for control technology development as well as health effects studies and monitoring.

In making any **field survey** and follow-up sampling the following questions must be answered:

- Where to sample?

- Whom to sample?

- How many samples to take?

- When to sample?

Additionally, the use of instruments requires:

- Accurate calibration and periodic recalibrations

- Flow and volume accuracy

- Calibration of collection efficiency

- Establishing sample stability and recovery

- Calibration of sensor response

- Use of reliable standards and standard procedures.

Process and workroom atmospheres must be sampled for the following reasons:

- To determine the nature and quantity of contaminants and to insure that levels are within acceptable and legal standards.

- To evaluate efficiencies of ventilation systems and/or environmental control equipment.

- For the selection and design of equipment.

- To project any potential problems in additions of new equipment.

- As evidence for regulatory compliance and for maintenance of safe working conditions for employees.

BIOLOGICAL AND INFECTIOUS HAZARDS

Fundamental to biological or **infectious material** risks of exposure is the proper program based on known identification and characteristics. The types of operations that potentially offer such risks are typically the **health care industry**, academic and industrial-research laboratories, the pharmaceutical industry, veterinary facilities, and food, drug, and cosmetic industries. Any operating system handling such materials should address:

- materials handling

- storage (when required)

- packaging for safe transport and effective treatment

- transport both in a facility and offsite

- appropriate treatment and disposal methods for by-products or wastes

- monitoring

- compliance with all ordinances and regulations.

The optimal system will vary from facility to facility but should address the above listed elements. Areas of concern not only involve the operations themselves but the handling and disposal of wastes from such sources.

Defining an infectious material has been ambiguous to say the least since different perspectives influence the criteria and even the terminology is not the same - infectious, pathological, biomedical, biohazardous, toxic, medically hazardous have all been used at times to describe similar materials and hazards.

The principal factors that are necessary to induce disease include:

- Presence of a pathogen

- Presence of a susceptible host

- A route of exposure for transmission of the pathogen to the host

- Exposure to a virulent pathogen

- Exposure to an infective dose.

In order for a material to become infectious, it must contain pathogens with sufficient virulence in adequate numbers to provide an infective dose. Disease induction requires a susceptible host. For a host to be susceptible, a variety of factors exist which include age, the person's state of health, general immune state, and the degree of immunity to a particular pathogen. Those that are usually very susceptible are the chronically ill, the very old, the very young, and the immune deficient.

The principle **routes of transmission** that are relevant to infectious materials as a source of disease are inhalation, ingestion, and percutaneous transfer. This does not mean that all exposure routes are necessarily related to disease induction; some pathogens are harmless or are killed in the digestive system, whereas if they enter the respiratory system they are pathogenic. Ingestion of pathogens can result from eating contaminated food products, or in the case of infectious wastes, ingestion could occur from hand to mouth when the hands become contaminated from handling contaminated material.

Types of infectious wastes include:

- Isolation wastes

- Cultures and stocks of etiologic agents

- Blood and its products

- Pathological wastes ("not necessarily always infectious")

- Sharps (used hypodermic needles) ("not necessarily always infectious")

- Other wastes from surgery, autopsy, dissection

- Contaminated laboratory wastes

- Dialysis unit wastes

- Animal carcasses and body parts

- Animal bedding and other wastes from animal rooms

- Discarded biologicals

- Contaminated food and other products

- Contaminated equipment.

Certain of these wastes as indicated are not necessarily always infectious but are included because they should be managed or handled to minimize the hazards and special problems of such materials.

Infectious materials pose a possible hazard not only to individuals in the occupational environment but also to overall human health and the public at large. Effective safety management practices from product/ process/activity/waste disposal practices are required. A safety program should be a comprehensive plan, in writing, encompassing all safety aspects for different and all types of materials - wastes including infectious, radioactive, chemical, general, as well as materials with multiple hazards (e.g., infectious-radioactive; infectious-toxic; infectious-radioactive-carcinogenic). It is appropriate for each laboratory or department to have specific, detailed, written instructions for proper safety management.

COMPRESSED GASES AND FLAMMABLES

Compressed gases and flammable liquids present special hazards and risks. Laboratories, manufacturing operations, gas welding, and other areas using flammable materials, where such gases and liquids are stored are considered hazardous.

Compressed gases and flammable liquids must be handled and stored properly to prevent spills or vapors from causing fires or explosions. Relatively small amounts of gases or vapors can ignite from an arc caused by an electric switch, cigarette, open flame, or other ignition source. Vapors can even travel considerable distances to ignition sources and flash back to the original source, causing an explosion. Areas where compressed gases and flammable liquids are used must be properly designed and well-maintained. These areas must also have specific safety features such as eyewashes, safety showers, and fire blankets. Fire protection features, such as fire-resistant construction and fire extinguishing systems, will further reduce the hazard potential.

HVAC

The **heating, ventilating and air conditioning (HVAC)** system is an important component in safety and fire protection. Both internal and external disaster plans must include provisions for HVAC systems. The needs of the HVAC system should be determined by administration, engineering staff, department heads, staff representatives and infection control personnel.

LABORATORY GASES

Many of the gases used in laboratories are toxic, explosive and corrosive. Large **cylinders of gas** can weigh over 100 pounds and small cylinders weigh approximately 25 pounds. They all contain gas that is under enough pressure to propel the cylinder like a missile if the valve is broken. Special precautions must be taken to assure that no slow leaks exist, since even small leaks can cause an explosion or asphyxiation. Particular attention is necessary when cylinders are placed near potential ignition sources.

Spare tanks should never be stored in use areas and replacement tanks should be brought to the laboratory only when an empty tank is removed. Whenever possible, large tanks should not be used in the laboratory. Some laboratory gases, however, such as acetylene used for atomic absorption, must be dispensed from larger tanks. These gases should either be dispensed from smaller tanks or piped into the laboratory.

If a regulator is to be used for a different type of gas, other than those for which it is intended, check with the regulator manufacturer to be sure that the materials in the regulator are compatible with the new gas. Never use adapters to adapt a regulator to other gases. Teflon tape or similar sealants cannot be used to assure a tight seal since these sealants will not stand up to pressure properly. The valve connection is designed for a correct metal to metal seal only. If additional protection is needed because a connector is damaged or worn or because it is the wrong connector, the valve should be replaced.

Though most gases have a long shelf life, tanks should be tested every five years. If a tank is needed for a long period of time or has a very low use application, smaller tanks or lecture bottles should be used. Tanks should be checked annually for leaks or other defects and any tank over five years old should be returned to the supplier for hydrostatic testing. Tanks should be returned to suppliers as soon as they are empty. Laboratory tanks should have a test date stamped on the upper surface, near the valve. These tanks should be hydrotested every five years and should not be kept beyond the fifth year after the last test date.

COMPRESSED GAS STORAGE

Compressed gases other than cylinders actually in use, should be stored in designated storage areas reserved for the storage of such gases. These areas should be ventilated by either mechanical or natural methods. **Oxidizers**, such as oxygen and nitrous oxide, must be separated from flammable gases, such as methane and ethylene oxide. Tanks in storage areas should be secured to the wall or to a similar secure structure, and should have their protective caps securely in place. All tanks should be stored upright.

Flammable liquids and compressed gases should be stored in different storage areas. No flammable materials of any kind can be kept in a compressed gas storage area, with the exception of packaging used to transport and separate small tanks. Such combustible materials would add to the fire load and, in case of a fire, could melt the cylinder safety disk, release the gas, and cause a blowtorch effect. For example, even though oxygen does not burn, it can cause a fire to burn more intensely.

Small cylinders should be stored in their shipping boxes or stored in racks, shelves, or bins that preferably are non-combustible. Small cylinders should never be chained or strapped to walls or sturdy structures, but do not require special racks. Since cylinders are not stable, they should not sit upright without some method of secure support. Cylinder storage areas may be vented mechanically, by using fans that meet the requirements for Class I, Division 2 motors outlined in the National Electrical Code. The system must vent to the outside and must vent at a sufficient distance from air intake systems to preclude recirculation of any gases.

Separate storage rooms for empty tanks are not required. Empty tanks may be stored in the same room as full ones, however, tanks should be well marked and in different groups. Tanks should always be stored and handled with protective caps in place. The caps should be removed just before use and replaced when use is discontinued or the tanks are empty.

HANDLING AND STORAGE OF FLAMMABLE LIQUIDS

Improper handling and use of flammable liquids is the primary cause of fires. Where possible, flammable liquids should not be used or be stored near areas of ignition, such as electrical switches that may spark, open flame, heating devices, or other heat sources. Since vapors from flammable liquids are heavier than air, there is an additional danger of them traveling along the floor to other ignition sources. Flammable liquids should be handled in hoods or well-ventilated areas over a spill tray to contain small incidental spills. Small spills should be picked up and not allowed to vaporize.

Highly flammable materials, like ether or acetone, should be purchased in the smallest practical container to minimize spill, waste, and the need for handling. Bench containers or dispensers should contain no more than a two-day supply, and should be replenished only as needed.

Glass may be used to contain small amounts of flammables. Highly flammable materials should not be in glass bottles larger than a pint. Less flammable materials can be in liter or quart bottles. Whenever possible, metal cans should be used to reduce breakage. Because of the breakage factor, plastic bottles are also safer than glass. In the case of reagents that are purchased in gallon-sized bottles, these must be transferred to small dispensers to reduce handling and spillage.

Label information should include the name and formula of the material; its hazard class; receipt, opening, and expiration dates; and whenever applicable, emergency instructions. Additional information should be included for highly flammable, toxic, carcinogenic, and corrosive flammables, or those that may react with other chemicals in an uncontrolled reaction if spilled or broken.

A small spill should never be allowed to vaporize. Such spills should be absorbed with absorbent granulated charcoal or, if nothing else is available, paper towels. The absorbent and liquid should be put in a sealable bag or container, and handled as a flammable waste. Whenever any spill occurs, all ignition sources must be eliminated until the ventilation system can clear the air.

Safety cans have spring closing tops to minimize spills if dropped. Such a top also allows controlled venting and prevents rupturing if the can is exposed to high heat. If the spring top is a nuisance, there are safety cans available with pouring spouts and dispensing valves. The metal mesh is a flame arrestor used to keep a flame front from entering the safety can. The metal screen cools the flame and puts it out. The mesh, of course, must be cleaned whenever it becomes dirty.

BULK FLAMMABLE LIQUIDS STORAGE

Flammable liquids are widely used and bulk supplies should be stored away from operations in a special flammable liquids storage vault or in flammable liquids storage tanks in sprinklered areas. Quantities of more than 60 gallons must be stored in vaults or in one hour fire-resistive rooms equipped with sprinklers. Amounts below 60 gallons can be kept in flammable liquids storage cabinets until dispensed for use.

For flammable liquids that are dispensed from drums, the drums should be upright, and liquids should be dispensed by means of hand pumps with drip pans. The drums should be properly grounded with ground wiring, and metal safety cans filled from drums should be electrically bonded to the drum with an alligator clip and wire. Smoking prohibition must be strictly enforced and proper ventilation, exhaust to the outside with no recirculation, is necessary in all flammable liquids storage areas. All electrical equipment must meet the requirements outlined in the National Electrical Code, 1981 edition, for Class I, Division 1 explosion-proof equipment.

Highly flammable liquids, such as ether and acetone, should not be poured or pumped from large cans to small bottles. Because of the risk of fire and explosion, such liquids should be purchased in containers of a size appropriate for their use. When dispensing is unavoidable, it should be done in a chemical fume hood to avoid potential ignition.

IGNITION SOURCES

Smoking, matches, lighters should not be allowed in flammable areas. Such areas are not difficult to police; smoke breaks and substitutes such as chewing gum should be allowed to relieve employee stress.

Static electricity is concerned with electricity generated by people's clothing, shoes, work processes. All filtering operations and equipment should be grounded to drain off electrical charge.

Electrical equipment - the turning on/off of light switches, making and braking contactors in electrical control equipment and heat generated by hot wiring motors, all present potential ignition sources capable of causing an explosion.

Friction - gears, belting and mechanical transmission equipment improperly maintained produce friction heating. Dusts may combine with loose grease or improperly sealed equipment to break down the lubricant and due to added dust ignite upper lubricant layers.

Mechanical sparks - non-sparking materials should be used wherever possible. Scoops to handle wet-dry materials, clean tanks, and fan housings. Magnetic metal separators and gravity traps should be installed for tramp metal removal in conveyor and process areas, milling/grinding and related equipment areas.

Hot work permits should be secured anytime welding, grinding, and equipment repair takes place. Flames, hot surfaces, and sparks can provide ignition sources for flammable liquids, dusts, and fibers.

Spontaneous ignition is possible in processes handling fine powders, oil saturated products. Such operations should have temperature detectors and alarms to monitor abnormal temperature rises.

Reactivity data should be given on OSHA 20 **Material Data Sheet** indicating degree and conditions of decomposition, polymerization, incompatibility with other materials.

EXPLOSIVE OR FLAMMABLE MATERIALS

Explosive or flammable materials can be divided into three groups as corresponding to the National Electrical Code:

- Class I (liquids and gases)
- Class II (dusts)
- Class III (fibers)

In considering flammable liquids and gases the most significant properties are: boiling point; flash point; evaporation rate; specific gravity; vapor volume; and explosive limits.

Boiling point is the point at which a liquid's vapor pressure is equal to the atmospheric pressure at its surface. The boiling point of a material is the temperature at which this takes place.

Flash point is the minimum temperature at which a liquid gives off vapor with a test vessel in sufficient concentration to form an ignitable mixture with air near the surface of the liquid, and can be determined by two ASTM tests (ASTM D-56-70 Flash Point by Tag Closed Tester/STM D93-71 Flash Point by Pensky-Matens Closed Tester).

Vapor Volume is the number of cubic feet of solvent vapor formed by the vaporation of one gallon of solvent.

Evaporation rate is the process by which a liquid is changed to other gaseous state and mixes with surrounding air or vapors. Evaporation rate is the length of time required for a given amount of material to evaporate.

Specific gravity - ratio of the mass of a given volume of liquid to the mass of an equal volume of water.

Flammable limits (Explosive limits) - lower flammable limit (LFL) or lower explosive limit (LEL) is the lowest concentration of a combustible or flammable gas or vapor in air that will produce a flash of fire. Mixtures below this concentration are too "lean" to burn. Upper flammable limit (UFL) or upper explosive limit (UEL) is the highest concentration of a combustible or flammable gas or vapor in air that will produce a flash of fire. Mixtures above this concentration are too rich to burn (or oxygen deficient).

Autoignition temperature is the lowest temperature at which a material begins to self-heat at a high enough rate to result in combustion.

ELECTRICAL EQUIPMENT HAZARDS IN HAZARDOUS ENVIRONMENTS

Explosion-proof equipment or apparatus is that which is encased or enclosed in such a manner that it is capable of withstanding an explosion of a specified gas or vapor which may occur within it and of preventing the ignition of a specified gas or vapor surrounding the enclosure by sparks, flashes, or explosion of the gas/vapor within. Operation is at such external temperature that a surrounding flammable atmosphere will not be ignited.

Dust-ignition proof equipment is enclosed in a manner that will exclude ignitable amounts of dust or amounts that might affect performance or rating and that when installed in accordance with appropriate codes, will not permit arc, sparks, or liberated heat generation inside the enclosure or to cause ignition of exterior accumulations of atmospheric suspensions of a specified dust.

The **National Electric Code** is widely used for classification purposes and divides atmosphere explosion hazards into three broad classes:

Class I - gases, vapors Division 1 normally hazardous

Division 2 not normally hazardous

Class II - Combustible dusts Division 1 normally hazards

Division 2 not normally hazardous

Class III - Easily ignitable Division 1 and 2 - covers lighting fibers and fixtures

Class I atmospheric hazards are divided into groups as well as divisions. Division I covers locations where flammable gases or vapors may exist under normal operating conditions, under frequent repair or maintenance, or where breakdowns or faulty operation of process equipment might also cause simultaneous failure of electrical equipment. Division 2 covers locations where flammable gases, vapors, or volatile liquids are handled in a closed system; confined within suitable enclosures or where hazardous concentrations are controlled by positive mechanical ventilation. Areas adjacent to Division 1 locations into which gases might occasionally flow, would belong to Division 2.

Class II atmospheric hazards cover three groups of combustible dusts. Groupings are based primarily on dust resistivity and to some degree on temperature.

Class III atmospheric hazards cover combustible fibers or flyings not likely to be in suspension in air in sufficient quantities to produce ignitable mixtures (Division I) and locations in which easily ignitable fibers are stored or handled (Division 2).

For more details one should refer to the National Electric Code and Underwriters Laboratories Inc.

Hazard types and potential causes for fire include:

Possible Causes:

Open flame
Matches/smoking

Nearby fires
Fired process equipment
Gas heaters
Welding/flame cutting
Sparks
Static discharges
Electrical equipment
Mechanical (hot solids)
Chemical (carbon particles)
Combustible mix at auotignition
External heat
Electrical heaters
High wattage electronics
Boilers/radiators/steam lines
Exhaust manifolds
Hot process equipment
Friction
Spontaneous ignition
Oily rags
Sawdust
Powdered plastics
Gas mix at adiabatic compression
Pyrophoric reactions
Reactions with water sensitive materials

Hazard types and potential causes for explosions

Possible Causes:

Explosives
Combustible gases in confined spaces
Fine dusts and powders
Combustible gases or liquids
Strong oxidizers
High temperatures
Cryogenic or highly volatile fluids heated
Fuel/lubricant/solvent in contact with strong oxidizer
Delayed combustion in firing chamber
Hydrogen ignition from battery charging
Water/moisture contact with sensitive materials (sodium, potassium, lithium)

RESPIRATORY HAZARDS

Respiratory hazards exist if a particular airborne material is sufficient toxic or present in such concentration to effect damage to the worker in his environment or have cumulative or residual effects. The route of entry to the body is by inhalation for such a hazard. The respiratory system conveys air (oxygen) to the lungs where oxygen is transferred to the blood which is carried through the circulatory system to cells in the human body. Toxic materials in the air follow this route into the body and can enter via one of three methods:

- inhalation through the respiratory tract,

- eating and swallowing through the digestive tract,

- through the skin by absorption.

The most important route and greatest in risk potential in the industrial environment is the respiratory tract. Respiratory hazards manifest themselves as either:

- nuisance atmospheres

- toxic atmospheres

- oxygen deficient atmospheres.

Toxic and nuisance atmospheres in the occupational environment may be gases, liquid particles, or solids. **Irritants** typically inflame moist body surfaces (eyes, nose, throat, mouth) and the type, concentration, dose will determine the severity of the irritation. Suspended liquid aerosols and particulates as well as gaseous contaminants are the physical manifestations of the risk.

Dusts, fumes and mists in general, particle sizes and settling rate are the most important characteristic of airborne suspended matter. Particles larger than 100 microns (μ) may be excluded from the category of dispersions because they settle rapidly. Particles of one micron or less in size will settle at slower rates, and are thus usually considered to be permanent suspensions. A thorough understanding and examination of the occupational atmospheric environment must be made.

Dusts are formed by such operations as pulverization of solid matter into small size particles. These may be the results of such processes as grinding, crushing, blasting, drilling. Particle size range from 1 to μ up to 100-200 μ Dust particles are usually irregular in shape and heterogeneous in size and structure.

Smoke includes the derivatives from combustion of organic matter (oil, coal, wood, tobacco), gaseous particle suspensions by chemical and photochemical reactions, condensations, and volatilizations. Smoke particles are extremely small, with sizes ranging from less than 0.01 μ to 1 μ. They are usually spherical in shape if of liquid or tarry composition and if they are of solid composition, an irregular shape is characteristic. Smog refers to a mixture of material fog (water condensation) and industrial smoke.

Fumes are formed by processes such as sublimation, condensation, and combustion. A common type is generated from oxidation of metallic vapors or compounds. Smelting operations for example generate metallurgical fumes of this type such as oxides of zinc, cadmium and beryllium. Particle sizes range from 0.1μ to 1μ.

Mists are formed by the condensation of vapors upon a suitable nucleus which gives rise to a suspension of small liquid droplets. Particle sizes of naturally occurring mists range between 5μ and 100μ and submicron particles can be produced under special conditions. Terminology here is used rather loosely. For example, some industrial dispersoids such as sulfuric acid particulates are termed mists, even though they are one micron or less in size, which would put them in the category of smoke or fumes.

Gases or gaseous contaminants are a form of matter having extreme molecular mobility and capable of diffusing and expanding rapidly in all directions. The individual size is at the molecular level. The level of risk and toxicity depends upon the chemical nature, concentration, and duration of exposure.

Systemic reactions can produce lung damage of various types caused by toxic dusts. These can result from exposure to specific organic compounds or heavy metals (cadmium, mercury, lead, manganese, etc.) For example, emphysema and pneumonitis can be affected by exposure to cadmium.

Metal fumes exposure by inhaling very fine, newly produced zinc and/or magnesium or their oxides produce metal fume fever.

Allergic reactions and **irritation/sensitization reactions** are caused by skin contact or inhalation of fine organic dusts. Irritation of the nose and throat is caused by acid or alkali dusts or mists. Certain materials in this class such as chromates can result in ulceration of nasal passages and even lung cancer.

Asphyxiants interfere with the oxygen use at the tissues. Simple asphyxiants are physiologically inert gases that simply act to dilute the oxygen if the air causing suffocation; these include carbon dioxide, methane, ethane, nitrous oxide. Chemical asphyxiants via chemical action prevent transporting oxygen to the blood/circulatory system from the lungs or inhibit oxygen release to the tissue cells. Such compounds include carbon monoxide, cyanogen, nitrites, aniline and its derivatives, hydrogen sulfide. Carbon monoxide is a classic example of asphyxiant action. Carbon monoxide has an affinity for the hemoglobin of the blood 210 times greater than oxygen, to form carboxyhemoglobin which prevents oxygen from being transported to cells and tissues.

Anesthetics and narcotics act as **depressants** on the central nervous system, particularly the brain. Some examples in this group are: acetylene hydrocarbons; olefin hydrocarbons; ethyl and isopropyl ether; paraffin hydrocarbons; aliphatic ketones; aliphatic alcohols; esters which hydrolyze in the body to form organic acids; and alcohols. These materials on inhalation pass into the blood stream and are carried to the central nervous system acting as depressants. Symptoms include mental confusion and drowsiness.

Systemic poisons are materials which as their main toxic action produce damage to the central nervous system, the blood system, and organs including the liver and kidneys. Hepatotoxic agents produce liver damage; an example is carbon tetrachloride, producing severe diffuse central necrosis of the liver. Tetrachloroethane produces acute yellow atrophy of the liver and nitrosamines are capable of severe liver damage. Nephrotoxic agents effects produce kidney damage; included in this class are halogenated hydrocarbons which can produce liver damage also. Neurotoxic agents are materials whose toxic effects can produce nervous system damage. Examples are metals such as mercury, thallium, and manganese. The central nervous system is sensitive to neurological damage by organometallic compounds such as methyl mercury, tetraethyl lead.

Blood or hematopoietic system agents such as nitrates, aniline, toluene convert hemoglobin to methemoglobin. Nitrobenzene lowers blood pressure as well as forming methemoglobin; arsine produces hemolysis of red blood cells, benzene damages hematopoetic cells of bone marrow.

The resulting biological or physiobiological reaction in relation to toxic air contaminants depends upon the concentration and duration of exposure. Of the classes described it may frequently not be possible to place a substance within a single class of physiological classification. Work areas must be evaluated to determine whether respiratory hazards exist as well as whether existing engineering or administration controls are protecting the worker.

The fundamental remise in establishing any **respirator program** is that accepted engineering controls are not feasible, or for use on an intermittent-standby-emergency basis. Until the enactment of the Occupational Safety and Health Act (OSHA) in 1970, most guidance on respiratory protective devices, (respirators) use in hazardous environments was advisory rather than mandatory. Now, OSHA Part 1910.134 and 1910.146 set forth specific legal requirements for selection, use, and maintenance of respirators, and establish guidelines for a respirator program to meet those requirements.

INFORMATION AND TRAINING PROGRAMS REQUIRED BY OSHA

As part of the safety and health program, employers are required to develop and implement a program to inform workers (including contractors and subcontractors) performing hazardous waste operations of the level and degree of exposure they are likely to encounter.

Employers also are required to develop and implement procedures for introducing effective new technologies that provide improved worker protection in hazardous waste operations. Examples include foams, absorbents, adsorbents, neutralizers, etc.

Training makes workers aware of the potential hazards they may encounter and provide the necessary knowledge and skills to perform their work with minimal risk to their safety and health. The employer must develop a training program for all employees exposed to safety and health hazards during hazardous waste operations. Both supervisors and workers must be trained to recognize hazards and to prevent them; to select, care for and use respirators properly as well as other types of personal protective equipment; to understand engineering controls and their use; to use proper decontamination procedures; to understand the emergency response plan, medical surveillance requirements, confined space entry procedures, spill containment program, and any appropriate work practices. Workers also must know the names of personnel and their alternates responsible for site safety and health. The amount of instruction differs with the nature of the work operations.

Employees at all sites must not perform any hazardous waste operations unless they have been trained to the level required by their job function and responsibility and have been certified by their instructor as having completed the necessary training. All emergency responders must receive refresher training sufficient to maintain or demonstrate competency annually. Employee training requirements are further defined by the nature of the work (e.g., temporary emergency response personnel, firefighters, safety officers, HAZMAT personnel, incident commanders, etc.). These requirements may include recognizing and knowing the hazardous materials and their risks, knowing how to select and use appropriate personal protective equipment, and knowing the appropriate control, containment, or confinement procedures and how to implement them. The specific training and competency requirements for each personnel category are explained fully in the final rule (54 FR 42:9294, March 6, 1989). The following table can be used as a basis for establishing the type of training required at your facility.

HAZARDOUS WASTE/CHEMICAL WORKER TRAINING CHECKLIST

The following checklist is useful to environmental managers in meeting compliance issues. When you join NASHP, you receive helpful forms like these which help to identify your compliance needs.

OSHA'S 'HAZWOPPER' (29 CFR Part 1910.120)

	Yes	No
1. Do you employ any of the following types of workers:		
• Federal or state hazardous-waste site cleanup workers?	☐	☐
• RCRA corrective-action cleanup workers (hazardous waste site at a facility that is still operating)?	☐	☐
• Hazardous-waste workers at treatment, storage, and disposal facilities (TSDFs)--general laborers, equipment operators, their supervisors, on-site hazmat teams?	☐	☐
• Hazmat teams--including company hazmat teams and fire department teams?	☐	☐
• Basically, any workers who may be brought in from outside the immediate area of a spill to respond to it, even if they are not part of a formal hazmat team?	☐	☐

If you do *not* have *any* of these workers, go on to the next law.

2. If you **do**, then have they received *at least* the amounts of training indicated?
   - Hazardous waste cleanup sites (including at state/federal Superfund sites and RCRA corrective action cleanups)--

	Yes	No
Hazwaste cleanup workers: *40 hours off-site, 3-day on-the-job, and annual 8-hour refresher courses*	☐	☐
Supervisors of cleanup workers: *Same as the workers, plus 8 hours of "specialized training*	☐	☐
"Occasional" and "low-hazard" workers: *24 hours off-site, 1 day on-the-job, 8-hour refreshers*	☐	☐
Emergency responders at these sites: training amounts not specified, but need *training prior to response work; annual rehearsals of emergency-response plan*	☐	☐

   - Hazardous waste TSDFs--

	Yes	No
Hazardous waste workers: *24 hours of training, 8-hour refreshers*	☐	☐
Emergency responders: *Amount* of training not specified by OSHA, which instead describes the required *level* of training in the following manner: *"As necessary, prior to responding to emergencies, to develop competency to protect themselves and others."*	☐	☐

   - Hazmat teams and other emergency responders (five different training levels, increasing in expertise as the level number increases):

	Yes	No
1. First responder awareness level: *no specified hours for initial training, but enough to recognize an emergency and report it to authorities; 8-hour yearly refresher training*	☐	☐
2. First responder operations level: *8 hours of initial training and 8 hours annual refresher*	☐	☐
3. Hazardous materials technician: *24 hours off-site and 8-hours annual refresher*	☐	☐
4. Hazardous materials specialist: *24 hours of training off-site; 8 hours yearly refresher*	☐	☐
5. On-scene incident commander: *24 hours of off-site training; 8-hours annual refresher*	☐	☐

3. Have all of these workers been trained in the following general areas (1910.120(e)):

	Yes	No
• The names of personnel and alternates who are responsible for site safety and health?	☐	☐
• The safety hazards that are present at the site?	☐	☐
• Proper use of personal protective equipment?	☐	☐
• Work practices by which employees can minimize hazard risks?	☐	☐
• Safe use of engineering controls and equipment at the site?	☐	☐
• Medical surveillance requirements, including the recognition of symptoms and signs that may indicate a chemical overexposure?	☐	☐
• Site control measures?	☐	☐
• Decontamination procedures?	☐	☐
• Site's standard operating procedures?	☐	☐
• Site's contingency plan?	☐	☐
• Confined-space entry procedures?	☐	☐

	Yes	No
4. Have you made sure that any worker that is not "certified" under this provision, or has not been "grandfathered" in under 1910.120(e)(9), does [not] engage in hazardous waste work?	☐	☐

OSHA's Hazard Communication Standard (29 CFR Part 1910.1200)

	Yes	No
5. Do you employ any worker who may be exposed to hazardous chemicals--such as the numerous substances listed under Part 1910, Subpart Z (excluding RCRA hazardous waste)--not counting nonroutine, isolated instances? **If *no*, go on to the next law.**	☐	☐

6. If *yes*, then have you trained them in the following areas (1910.120(h)(2)):

	Yes	No
• The methods of detecting the presence or release of a hazardous chemical in the work area (e.g., employer-conducted monitoring; continuous chemical-monitoring devices; visual appearance or odor of the chemical when released?	☐	☐
• Physical and health hazards of the chemical(s) used?	☐	☐
• Measures workers can take to protect themselves against dangerous exposures, including emergency procedures and personal protective equipment?	☐	☐
• Details of the HCS program, including explanations of the material safety data sheet (MSDS) and container-labeling system, and how they can obtain and use this information?	☐	☐

7. Have you double-checked whether the following types of workers are adequately trained under HCS (1910.1200(h)):

	Yes	No
• Any worker that has been *newly assigned to a task that falls under the HCS?*	☐	☐
• Any worker around a *new hazard that has been introduced into the workplace?*	☐	☐

OSHA's Process Safety Standard (29 CFR Part 1910.119)

8. Do you have any processes that involve any of the following:

	Yes	No
• Use of a "highly hazardous chemical" above its threshold quantity (e.g., a process that uses 2,500 pounds of chlorine);	☐	☐
• 10,000 pounds or more of flammable liquids or gases (except for hydrocarbon fuels used solely for workplace consumption and flammable liquids that are stored below their normal boiling point, without the benefit of chilling or refrigeration);	☐	☐
• Explosives or pyrotechnics manufacture? **If *no*, go on to the next law.**	☐	☐

9. If yes, do you fall under any of the law's exemptions, as follows:

	Yes	No
• Retail facilities?	☐	☐
• Oil or gas well drilling or servicing operations?	☐	☐
• Normally unoccupied remote facilities? **If *yes*, go on to the next law.**	☐	☐

	Yes	No
10. If *no*, then have you provided "each worker operating a process, or who is newly assigned to a process? with training in that process's operating procedures (1910.119(g))?	☐	☐
11. Does this training emphasize safety and health hazards; emergency operations including shutdown, and other safe work practices?	☐	☐
12. Does your process safety training also include:		
• The mechanical integrity of the process (1910.119(j)(3))?	☐	☐
• The ramifications of any change in the process, prior to start-up (1910.119(l)(3))?	☐	☐
13. Have you evaluated any contractor's training responsibilities under this standard (1910.119(h)(2))?	☐	☐
14. Have contractors given you some assurance that each contract worker is trained in his/her safe work practices (1910.119(h)(3)(v))?	☐	☐

CHAPTER 9

ESTIMATING RELEASES TO THE ENVIRONMENT

SOURCES OF WASTES AND RELEASES

All sources of wastes should be considered in estimating releases of a chemical from a facility. Sources include but are not limited to the following:

Fugitive air sources

- Volatilization from open vessels, waste-treatment facilities, spills, and/or shipping containers

- Leaks from pumps, valves, and/or flanges

- Building ventilation systems

Stack or point air sources

- Vents from reactors and other process vessels

- Storage tank vents

- Stacks or vents from pollution control devices, incinerators, etc.

Water sources

- Process steps

- Pollution control devices

- Washings from vessels, containers, etc.

- Storm water (if your permit includes storm water sources of a listed chemical)

Solids, slurries, and nonaqueous liquid sources

- Filter cakes, and/or filter media

- Distillation fractions

- Pollution control wastes such as baghouse particulates, absorber sludges, spent activated carbon, and/or wastewater treatment sludge

- Spent catalysts

- Vessel or tank residues (if not included under water sources)

- Spills and sweepings

- Off-specification product

- Spent solvents

- Byproducts

Accidental or nonroutine releases should also be included in the release totals, and are not to be listed separately. For example, fugitive air emissions estimated separately for leaks, open vessels, and spills would be considered under "Fugitive or Nonpoint Air Emissions."

So that consideration of all the possible points/sources of release is ensured prior to making the release estimate, it will be useful to prepare or refer to simplified flow diagrams for those processes involving specific pollutants; for example, for a process that uses a specific chemical, a schematic of the major pieces of equipment in which the process is carried out, the associated storage vessels, and the treatment steps for wastes containing the solvent would be helpful in assessing possible release points/sources. If the chemical is made or used in multiple processes, the quantities to be reported are the total releases for all processes; a flow diagram for each process would also be helpful.

AN OVERVIEW OF THE ANALYSIS

The level of detail of the analysis and the level of effort required depend on specific circumstances. Before data needs are described and before methods are outlined for estimating quantities, it should be noted that many (if not most) processors and users will have only one or two releases of a given chemical to report or be concerned with. Further, if monitoring data are available for that release, simple multiplication of the concentration of the chemical in the waste by the volume of the waste released may yield an acceptable estimate.

The following are examples of this "simple" solution:

- A furniture maker uses a listed solvent in coating furniture. The solvent evaporates in a drying area, from which it is ducted to a discharge stack and is then released into the air without treatment. In this case, the release estimate would simply be the amount of solvent present in the coating(s) purchased (adjusted for any inventory change). This value would be a point source emissions to air.

- A food processor uses an aqueous cleaning solution that contains a listed, nonvolatile component to wash down food processing equipment. In this case, the quantity of cleaning solution used multiplied by the concentration of the nonvolatile component in the cleaning solution would be used as an estimate of the release, say to a POTW (assuming that it does not undergo treatment prior to discharge). Recall that POTW means Publically Owned Treatment Works.

- The manufacture of a chemical compound in solution generates a solid filter cake that is land-filled on site. The filter cake contains a problem chemical. The release of the listed chemical would be estimated by multiplying the concentration of that chemical in the filter cake by the quantity of the filter cake landfilled in the reporting year. This estimate would then be categorized as a "release to land."

- A processor of copper-containing compounds has measured the concentration of copper in wastewater to comply with a water discharge permit. The copper concentration times the daily volume of wastewater times the number of days on which discharge occurs yields the release estimate. This estimate would be considered a "Discharge to Water."

In all of the above situations, readily available data on the volume of the chemical manufactured, processed, or used and data from the measurement of the concentration of the chemical in the waste were all that was needed to estimate a release. Of course, careful scrutiny of the process(es) at the facility is necessary to ensure that no sources are overlooked. For example, discarded containers of unused coating or water used to wash a filter press may be additional sources in the first and third examples, respectively.

The task will be somewhat more complicated when, for example, there are several waste streams, treatment is used, or wastewater is discharged but the chemical in the wastewater has not been measured. The following are examples of slightly more complex situations:

- A paint formulator incorporates a listed pigment into coatings. The formulator has determined that there are two sources of release for the listed pigment: 1) fine solids emitted to air from a milling step, and 2) solvent cleaning wastes that are sent to an off-site location for incineration. In this case, total release would be equal to the amount of pigment used (purchases adjusted for inventory changes) minus the amount of pigment sold in the product (the concentration of the pigment in the coating multiplied by the weight of coating solid). Because two wastes are involved, it is necessary to ap-

portion the total release between them. It is unlikely that "fugitive" solids to air will have been measured; therefore, the best approach may be to estimate the amount of cleaning waste (perhaps based on the known volume of the waste shipped offsite, the concentration of coating in the waste, and the concentration of the pigment in the coating). The release quantity in cleaning wastes calculated from these estimates would be considered as "Transfer to Offsite Location," and could then be subtracted from the total release estimate to yield the "fugitive air emissions."

- The processor of copper-containing compounds, discussed earlier, precipitates solids from wastewater generated by the process. In addition to the discharge mentioned previously, some precipitate is shipped to a waste broker. This additional copper release may be estimated by multiplying the volume of waste shipped by the concentration of copper in the waste. The type of disposal (transfer to a waste broker) would be indicated separately. Treatment efficiency may be calculated by dividing the amount of copper in solids by the total amount of copper (the amount of copper in solids plus the amount in the treated water). The resulting fraction would be multiplied by 100 to obtain a percentage reduction of copper in water resulting from the treatment (precipitation step). The concentration of copper in the influent would simply be the total copper in the two "releases" divided by the wastewater volume. (Alternatively, copper concentration in influent water may have been measured).

Calculations will be more complicated when a volatile material is made or used and air emissions must be estimated for leaks, vents, etc., or when no data are available on water releases and the water comes from several points in the process.

- The manufacture of a solvent uses a continuous process that involves a reactor, distillation columns, pumps, compressors, miles of piping, and hundreds of fittings as well as associated storage tanks and pollution control devices. Generally, the air release points will not have been monitored. Estimates of air releases must then be based on the other calculation techniques.

- The manufacture of a chemical generates wastewater during the reaction step. This wastewater is separated for treatment prior to discharge. Additional wastewaters arising from product washings and pollution control equipment are all combined in a central treatment system. The amount of chemical released can be estimated by considering the losses from each part of the process and then using mass balances and engineering calculations. Obviously, the larger the number of sources, the more difficult it will be to estimate the total release.

DEFINITIONS OF MAJOR APPROACHES

The preceding examples illustrated four basic approaches to estimating releases after

release points have been identified. These approaches are defined here:

- Calculations based on measured concentrations of the chemical in a waste stream and the volume/flow rate of that stream.

- Mass balance around entire processes or pieces of process equipment. The amount of a chemical leaving a vessel equals the amount entering. If input and output or "product" streams are known (based on measured values), a waste stream can be calculated as the difference between input and product (any accumulation/depletion of the chemical in the equipment, e.g., by reaction, must also be accounted for).

- Emission factors, which (usually) express releases as a ratio of amount released to process or equipment throughput. Emissions factors, which are commonly used for air emissions, are based on the average measured emissions at several facilities in the same industry.

- Engineering calculations and/or judgment based on physical/chemical properties and relationships such as the ideal gas law.

A single release estimate may involve the use of more than one of these estimation techniques; for example, when a mass balance is used to estimate the amount of wastewater leaving a process, and water solubility is used to calculate the maximum amount of chemical in that wastewater.

Estimates may be based on analogy. The emission factor approach relies heavily on your determination that your process is analogous to the process for which data were used to derive the factor. The use of any published data (for example, on the effectiveness of wastewater treatment for a chemical or on the releases from a papermaking plant) implies that the treatment schemes of processes are analogous to those your are using. Extreme caution should be used in the application of an analogy, especially from one facility to another.

OBSERVATIONS ON THE USE OF DATA

You may be able to estimate a release in several ways based on the various sets of data that are available. If this is the case, you will have to make a decision as to which estimate to report based on the expected accuracy of each. Assuming that equally valid and equally accurate data are available for each of the preceding approaches, the following caveats should be noted:

- Data on the actual released waste will generally provide a better estimate than data on the waste before treatment (to which a treatment efficiency must be applied).

- Data on the aggregate stream are preferable to data on the several streams that make up the aggregate.

- Data on the specific chemical are preferable to data on an analogue.

- Data on the chemical for a specific process are preferable to published data on similar processes. In fact, data on the treatment efficiency for a close analogue chemical treated at a

specific facility will probably provide a better estimate than published data on the actual chemical, as operating conditions vary greatly from plant to plant. It may be easier to make a good chemical analogy based on physical/chemical properties than to make a process analogy.

Data (for example, on the concentration of chemical in wastewater) may be available as a range of measured values. In this case, the average value of all measurements should be used for data specific to the facility as it operated in the reporting year, unless it can be demonstrated that some data points can be disregarded. If operating conditions varied during the year (e.g., the listed chemical was used periodically or new equipment was installed at midyear), releases should be estimated for each set of conditions (e.g., 3 months during which the chemical was used, 9 months during which it was not), and these values should be added. Representative data taken during a reporting period should be used. You should, however, consider whether including data from previous periods might improve the estimate because so few samples are taken each year.

With regard to published data on other processes, the average for facilities/ equipment/ operating conditions most closely analogous to the one in question should be used.

APPROACH TO USE

Selection of the best approach to estimating releases depends on the circumstances at your facility. Available information on a process may be the single most important factor in determining how to proceed. Provided are some general guidelines on the most effective approach(es), assuming that information is available to complete the analysis. It is organized according to type of release. There may be more than one approach.

Fugitive Air Emissions

Measurement data on fugitive air emissions will rarely be available. Furthermore, the fugitive emissions from most single sources is small compared with the total volume of chemical handled; therefore, inaccuracies in measurements of input and output can totally mask the magnitude of the release when the mass balance is attempted (an exception is the example of all solvent volatilized after application of a coating). For this reason, the use of emission factors is a major method for estimating fugitive air emissions. This approach requires the following:

- A published factor (usually reported as pounds emitted per pound of chemical processed or pounds emitted per piece of equipment, such as a valve).

- The amount of chemical handled at a facility and/or a count of the valves, pumps, etc., for which emission factors are available.

Specific emission factors are available for only a few processes as a whole, and these process-specific factors can only be applied to processes that are very similar to the one for which the factor was developed.

Volatilization equations can also be used for open vessels or for spills. This approach, however, requires that the vapor pressure of the chemical at the appropriate temperature, its molecular weight, and the open surface area be known or estimated.

Point Source Air Emission

Point-source air emissions or releases are much more likely to have been measured (as compared to fugitive air emissions). This permits calculations based on available data on the concentration and flow rate of the emission. For example, multiplication of the measured benzene concentrated by the measured flow rate of air through a vent yields the quantity of benzene being released. Unavailability of analytical techniques for determining airborne concentration of many of the chemicals limits this approach. When this is the case, a total hydrocarbon analysis can be used to establish an upper limit to the estimate.

Emission factors specific to some point sources (e.g., the reactor vent for ethylene dichloride production) are available and should be used if monitoring data are not available.

When these approaches are not possible, estimates for point sources must be based on mass balance calculations or on engineering calculations, design data, etc. Point sources such as storage tanks will usually require a calculation based on physical properties of the chemical, the throughput, and the configuration of the storage tank.

Releases to Wastewater

Many of the listed chemicals for a facility that may be subject to reporting requirements, are regulated under Federal, State, and/or local regulations. Frequently, wastewater discharges will have been monitored. If this is the case, release can be calculated directly. In fact discharge permit and Discharge Monitoring Reports may contain sufficient information to support any needed calculations (i.e., concentration of the listed chemical in the discharge and the wastewater flow rate). Multiplication of the measured concentration by the measured flow will yield an estimate of the release.

When monitoring data for a specific chemical not available at a facility, the following approaches may be applicable (in approximate order of preference):

- Identifying individual process points that contribute to water discharge, performing a mass balance calculation around each to determine individual releases, and then totaling them.

- Conducting a mass balance around the process as a whole. For example, input of dye equals output on dyed fabric plus output in wastewater (individual sources of that water need not be estimated). This approach is most appropriate if the only release of the listed chemical is through a wastewater stream.

- Using discharge data on the listed chemical from similar facilities. This approach is particularly useful if the industry has been studied by EPA's Office of Water Regulations and Standards and an Effluent Guidelines Background Document containing release estimates or typical waste stream concentrations for that industry is available.

Release in Solids, Slurries, and Nonaqueous Liquids

Some of these wastes may be regulated as hazardous wastes. Information in the permit and manifests for disposing of the waste provide a basis for estimating released quantities of a listed chemical. Frequently, however, the concentration of individual chemicals that make up a waste will not have been measured. In this case, the concentration of the chemical will have to be determined, either by measurement or by an estimation method based on mass balance, engineering calculations, etc.

For nonhazardous wastes in this category, the volume or total weight of the waste should be readily derivable from shipping records, a count of waste containers, etc. Again, the important factor to determine is the concentration of the listed chemical.

Unfortunately, there are no solid waste emission factors and little published data on concentrations of chemicals in such wastes. When monitoring data are not available for a waste, mass balance and engineering calculation approaches will be necessary.

ESTIMATING RELEASES TO AIR FROM STATIONARY SOURCES

Air emissions can originate from a wide variety of stationary sources and therefore are usually not centrally collected before being discharged; as a consequence, each source or category of sources must be evaluated individually to determine the amount released. Often, releases to air are reduced by the use of air pollution control devices, and the effectiveness of the control devices must be accounted for in the calculation of the release estimate. This section provides various methods for estimating releases to air and for determining the efficiency of pollution control devices.

Sources of Release to Air and Release Estimation Methods

Releases to air from industrial processes can be broadly categorized as follows: point sources, such as stacks and vents, and fugitive sources, which are not contained or ducted into the atmosphere. Whether a source is considered a point or fugitive source depends on whether the release is contained in a duct or stack before it enters the atmosphere. Common air emission sources that should be considered when estimating releases may be obtained from the literature. Examples illustrate the emission estimation methods for air emission sources. The examples presented are for purposes of illustration only; they are not meant to predict actual releases.

Process Vents

In general, process vents are the main air exhaust devices in a manufacturing or processing operation functioning under normal conditions; however, emergency venting devices on unit operations, such as relief valves, are also grouped under process vents. The methods that can be used to estimate releases to air from a process vent are discussed here; they include measurement, mass balance, emission factors, engineering calculations, or a combination of these methods. Several examples are given to illustrate the basic principles of each technique.

Measurement

Measurement is the most straightforward means of estimating releases . The pollutant concentration and flow rate from a process vent during typical operating conditions, if available, can be used to calculate releases. Total annual releases are based on the plant operating schedule for the year.

Example - Use of a Mass Balance to Estimate Air Emissions From a Process Vent:

Step 1. Draw a diagram, label all streams, and list input and output values

Consider a unit process that uses Chemical X to produce a product. In a year, 10,000 lb of Chemical X is used to produce 24,000 lb of a product containing 25 percent of Chemical X by weight. The input consists of 8000 lb of purchased Chemical X and 2000 lb that is collected from recycling. This process generates 5 tons or 10,000 lb of solid waste containing 15 percent (1500 lb) of Chemical X. The only other unit process stream is a process vent, which emits an unknown amount of Chemical X to the atmosphere. The following presents a schematic of this hypothtical unit process.

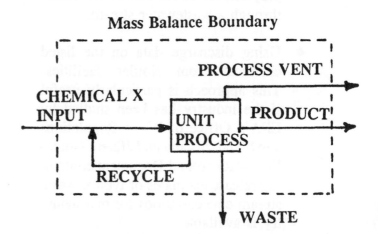

Hypothetical unit process using
Chemical X

Step 2. Set up equations with input streams equal to output streams.

Considering the quantities of Chemical X in all streams that enter or leave the process, the amount of Chemical X that is lost through the process vent on an annual basis can be estimated as follows:

Input = Amount purchased (8000 lb)

Output = Product (24,000 lb x 25%) + waste (10,000 lb x 15%) + process vent loss (unknown)

Input = Output

8000 lb Chemical X = 6000 lb + 1500 lb + process vent loss

Process vent loss = 8000 - 6000 - 1500 = 500 lb Chemical X per year

Example - Use of an Emission Factor to Estimate Toxic Air Emissions From a Process Vent:

Step 1. Assemble your emission factor information from the literature.

Hydrofluoric acid is being produced by reacting fluospar with sulfuric acid. The emission factor given in EPA Publication AP-42 is 50 pounds of fluoride per ton of acid product. The plant produced 55,000 tons of acid in the past year.

Step 2. Calculate releases.

In the absence of more accurate information, the **uncontrolled** fluoride emissions from the process would be calculated as follows:

$$\frac{55,000 \text{ tons}}{\text{year}} \times \frac{50 \text{ lb}}{\text{ton}} = 2,750,000 \text{ lb per year}$$

Based on information in AP-42, the use of a water scrubber to control releases would reduce emissions to 0.2 lb of fluoride per ton of acid. Emissions after control would thus be:

$$\frac{55,000 \text{ tons}}{\text{year}} \times \frac{0.2 \text{ lb}}{\text{ton}} = 11,000 \text{ lb per year}$$

Example - Use of Emission Factors to Determine Specific Chemical Air Emissions:

Step 1. Air emissions from the blast furnace of a primary lead smelting facility are controlled by a fabric filter system. In Section 7.6 of AP-42 (primary Lead Smelting), an emission factor for **uncontrolled** releases of particulate is given as 361 lb per ton of lead produced. Also in this section, a particulate removal efficiency range of 95 to 99 percent is provided for fabric filter control devices used for primary lead smelting operations.

Step 2. Calculate particulate releases.

Assuming the fabric filter system is 97 percent efficient, the particulate emission factor is reduced to

$$(1.00 - 0.97) \times \frac{361 \text{ lb particulate}}{\text{ton lead produced}} = \frac{10.83 \text{ lb particualte}}{\text{ton of lead}}$$

Thus, an annual production of 31,500 tons of lead will result in the emission of 341,000 lb of particulate (10.83 x 31,500).

Step 3. Calculate specific chemical releases.

The "Receptor Model Source Composition Library" is used to determine the amount of toxic compounds emitted. Source Profile No. 29302 gives a typical chemical composition for particulate matter sampled downstream of a fabric filter controlling emissions from a primary lead smelting blast furnace. Based on this information, annual emissions of individual toxic compounds can be calculated by multiplying the respective chemical composition by the total particulate 341,000 lb/yr. The specific compounds found according to this data source, their respective percentages of the total particulate matter, and their resultant annual emissions are summarized below.

Compound	Percentage of Particulate	Annual emissions, lb
Chromium	0.02	63
Nickel	0.06	189
Copper	0.35	1,197
Zinc	15.2	50,400
Cadmium	23.1	78,750
Lead	30.7	103,950

Example - Use of Engineering Calculations to Estimate Toxic Air Emissions from a Process Vent

A degreasing operation uses PCE to strip metal parts. A downstream solvent recovery operation is used to recover a portion of the PCE. For every pound of fresh PCE that is fed to the system, 0.2 lb of PCE is recycled. It is estimated that 0.5% of the PCE used in this process leaves with the product and is non-recoverable. Determine the PCE emissions that are vented through a process stuck at the degreaser tank.

Step 1. First construct a process diagram and label all streams.

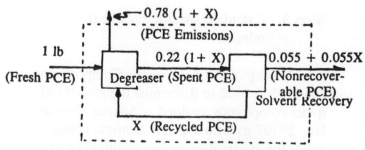

Step 2. Assign flow rates to each stream. As a basis of calculation, we can assume 1 lb of fresh PCE input. Therefore --

R = 0.2 lb PCE
Output
(Non-recoverable PCE) = 0.005 x (R + 1 lb PCE)
= 0.005 x (0.2 + 1)
= 0.006 lb PCE

Step 3. Now calculate the PCE Emissions from an Overal Mass Balance.

Input = Output
Fresh PCE In = PCE Emissions + Non-recoverable PCE
1 lb = PCE Emissions + 0.006
PCE Emissions = 0.994 lb

This value is in reality 0.994 lb of PCE emissions per pound of fresh PCE input, which was used as the basis for the calculations. Therefore, the total annual emissions of PCE would be 0.994 times the total amount of fresh PCE consumed annually.

Example - Calculate the Compositon of Vented Gas

A process vessel mixes three liquid components: A, B and C. The mixture contains 5 wt. percent A, 15 wt. percent B, and 80 wt. percent C, and vapors are vented to the atmosphere. The discharge rate through the vent has been measured at 5 ft^3 per minute at 70°F. The process vessel is in service 200 days per year. Determine the number of pounds of Chemical A that is emitted per year. The molecular weights of the chemicl ingredients are: A is 96, B is 360, C is 950.

Step 1. Calculate the composition of chemicals in the tank. Assuming equilibrium between the atmoshpere and conditions within the vessel, we can calculate the mole fraction of Chemical A liquid using the following formula:

$$X_{AL} = \text{Mole Fraction of Chemical A} = \cfrac{\cfrac{Wt.\ \%A}{MWA}}{\cfrac{Wt.\ \%\ A}{MW_A} + \cfrac{Wt.\ \%\ B}{MW_B} + \cfrac{Wt.\ \%\ C}{MW_C}}$$

where MW = molecular weight of compound

Assuming a 1 lb basis for calculations:

$$X_{AL} = \cfrac{\cfrac{0.05/96}{}}{\cfrac{0.05}{96} + \cfrac{0.15}{360} + \cfrac{0.80}{950}}$$

$$= \frac{5.21 \times 10^{-4}}{5.21 \times 10^{-4} + 4.17 \times 10^{-4} + 8.42 \times 10^{-4}} = 0.293$$

Step 2. Calculate the composition in the gas phase.

To do this we can use Raoult's Law:

$$P_A = X_{AL}P^o$$

where P_A = partial pressure of A at ambient conditions

P^o = vapor pressusre of A at ambient conditions

From a handbook of physical properties or an MSDS, we obtain the vapor pressure of A (25 mm pressure).

Therefore the partial pressure of A is

$$P_A = 0.293 \times 25\ mm = 7.33\ mm\ pressure$$

The fraction of A in gas is simply P_A/P_T, where P_T is the total pressure (1 atmosphere or 760 mm) or

$$X_{AG} = P_A/P_T = 7.33\ mm/760mm = 0.00964$$

Step 3. Calculate the annual release of Chemical A. This calculation can be made by use of the ideal gas law and recognizing that 1 lb-mole of gas occupies 359 ft^3 of volume. The following formula with appropriate process conditions inserted can be used:

$$X_{AG} \times 5\ \frac{ft^3}{min} \times \frac{60\ min}{h} \times \frac{24h}{day} \times 200\ \frac{operating\ days}{yr}$$

$$\times \frac{1\ lb\ mole}{359\ ft^3} \times \frac{\left(32°F + 460\right)°R}{70°F + 460\ °R} \times \frac{(MW_A)\ lb}{lb\text{-}mole} =$$

pounds of Chemical A emitted per year

$$Answer = 0.00964 \times 3723.6 \times 96 = 3446\ lb\ of$$
Chemical A emitted per year

Example - Use of an Emission Factor to Estimate Air Emissions From Material Storage:

Step 1. Assemble tank and product data.

The following calculations are for a 10,000-gallon, white, fixed-roof tank that holds, 1,1,1-trichloroethane at an average temperature of 60°F. The tank is 10 feet in diameter and 17 feet high. On the average, the tank is half full and has a throughput of 2000 gallons per month, or 24,000 gallons per year. The average diurnal (day and night) temperature change is 20°F. Ambient pressure is 1 atmosphere or 14.7 psi. Chemical handbook data show that 1,1,1-trichloroethane has a molecular weight of 133 and a vapor pressure of 1.6 psi at 60°F. The vapor pressure may be estimated by plotting temperature against vapor pressures obtained from handbooks and selecting the pressure at the given temperature.

Step 2. Insert values into equations and calculate releases.

Breathing losses (pounds/year)

0.0226	(factor)
x 133	(molecular weight)
x $\frac{1.6^{0.68}}{14.7 - 1.6}$	(vapor pressure ratio)
x $10^{1.73}$	(tank diameter)
x $8.5^{0.51}$	(half-full)
x $(20)^{0.5}$	(diurnal temperature change)
x 1	(paint factor for white)
x 0.51	(adjustment for small tanks)
x 1	(product factor)
= 262 pounds/year	

Working losses are estimated by use of the following equations developed by EPA for calculating loading losses for volatile organic liquids:

$$L_L = 12.46 \frac{SPM}{T}$$

where L_L = release in pounds/1000 gal of liquids loaded

P = liquid vapor pressure, psia (see chemical handbook or Appendix E)

M = molecular weight

T = liquid temperature (°F + 460)

S = Saturation factor depending on carrier and mode of operation as shown below:

Cargo carrier	Mode of operation	S factor
Tank trucks and tank cars clean	Submerged loading of a clean cargo tank	0.50
	Splash loading of a clean cargo tank	1.45
	Submerging loading: normal dedicated service	0.60
	Splash loading: dedicated vapor balance service	1.45
	Splash loading: dedicated vapor balance service	1.00
Marine vessels	Submerged loading: ships	0.2
	Submerged loading: barges	0.5

Working losses (pounds/year)

2.4×10^{-5} (factor)	
x 133	(molecular weight)
x 1.6	(vapor pressure)
x 10,000	(tank capacity)
x $\frac{24,000 \text{ gal used}}{10,000 \text{ gal capacity}}$	(turnovers per year)
x 1	(turnover factor)
x 1	(product factor)
= 123 pounds/year	

Total losses = 262 + 123 = 385 lb per year

The density of 1,1,1-trichloroethane is 11.2 pounds per gallon. Annual throughput is 24,000 gallons or 269,000 pounds. The calculated annual release is 385 pounds. A mass balance could not determine a 385 pound loss in 269,000 pounds handled.

Consequently, the use of emission factors is an appropriate method for estimating tank releases.

If the storage tank in the preceding example contained a mixture of materials A and B, the air releases could be calculated in a similar manner given the mole fractions of the components in the liquid phase (X_{AL} and X_{BL}) and the vapor pressure of the pure components (P^o and P^o). The molecular weight and vapor pressure used in the calculation of breathing and working losses would be calculated as:

$$\text{Molecular wt} = Mv = (M_A) \times \frac{P^o_x (A \quad AL)}{P^o_t} + (M_B) \times \frac{P^o_x (B \times BL)}{P^o_t}$$

$$\text{True vapor pressure} = P^o_t = (P^o_A)(X_{AL}) + (P^o_B)(X_{BL})$$

Example - Use of Measurement Data to Estimate Potential Toxic Air Emissions From Uncaptured Process Releases:

Step 1. Determine the basis for estimating releases and assemble necessary data.

Employee exposure to benzene should not exceed 1 ppm as an 8-hr time-weighted average. A plant has an alarm system that responds to 0.2 ppm benzene and a ventilation system that exhausts 20,000 acfm of room air at 70°F. If the alarm has not sounded during the course of the year and the plant operates 24 hours per day, 330 days per year, a conservative estimation of benzene fugitive releases could be performed as follows:

Step 2. Calculate releases.

Benzene releases per year would be calculated as follows:

$$\frac{20,000 \, ft^3}{minute} \times \frac{60 \, minutes}{hour} \times \frac{24 \, hr}{day} \times \frac{330 \, days}{year} \times$$

$$\frac{0.2 \, ft^3 \, benzene}{10^6 \, ft^3} \times = 1900 \, ft^3$$

The density of benzene vapor is 0.2 lb/ft³, and the annual release would be less than:

$$\frac{1900 \, ft^3 \, benzene}{year} \times \frac{0.2 \, lb}{ft^3} = 380 \, lb \text{ of benzene per year}$$

This value thus serves as an upper limit of potential releases.

Example - Use of Emission Factors to Estimate Toxic Air Emissions From Leaks in Vessels, Pipes, and Valves:

Step 1. Compile an inventory of fittings and appurtenaces that may leak organic compounds.

A chemical plant uses benzene (a light liquid with a vapor pressure greater than 2 psia) and has six pipe valves, three open-end valves, four flanges, two pumps, one compressor, and one pressure-relief valve. The plant operates 24 hours a day, 250 days a year.

Step 2. Review maintenance schedule and select appropriate emission factors based on leak rates.

The following calculation uses light liquid service factors and units of pounds per hour from Appendix G:

	(6 x 0.016)	(pipe valves)
+	(3 x 0.0037)	(open-end valves)
+	(4 x 0.0018)	(flanges)
+	(2 x 0.11)	(pumps)
+	(1 x 0.5)	(compressor in vapor service)
+	(1 x 0.23)	(pressure-relief valves in vapor service)
=	1.06 pounds per hour	

Average emission factors were obtained from the literature.

$$\frac{1.06 \text{ lb}}{\text{hour}} \times \frac{24 \text{ hr}}{\text{day}} \times \frac{250 \text{ days}}{\text{year}} = 6360 \text{ pounds per year}$$

ESTIMATING RELEASES IN WASTEWATER

At most facilities, wastewater from individual process sources is centrally collected and discharged from one point. This greatly simplifies the task of estimating releases of toxic materials to water because it decreases to one or a few the number of discharge streams for which releases must be estimated. Nevertheless, in some situations it may be necessary to estimate releases in wastewater from individual sources.

A facility that discharges or has the potential to discharge water containing toxic and/or hazardous wastes probably operates under the terms of Federal, State, and/or local permits. The permit(s) usually require measurements of the water volume and analyses of some generalized wastewater parameters [e.g., biological oxygen demand (BOD) and total suspended solids (TSS)]. Occasionally, releases for which the permit requirres analyses and those subject to reporting will be similar. In these instances, releases can be calculated by straightforward multiplication of the volume of wastewater released by the concentration of the chemical released. The permit(s) also often require that the wastewater be treated before its discharge to minimize releases.

The following subsections present some of the various sources of wastewater and methods of wastewater disposal. Also discussed are methods for calculating releases of compounds subject to reporting in wastewater and estimating efficiencies of wastewater treatment devices.

SOURCES OF WASTEWATER AND METHODS FOR ITS DISPOSAL

Releases of toxic chemicals can originate from a variety of wastewater sources. Some of the more common sources and process that generate wastewater is given. Unlike air emissions, wastewater from individual sources in a facility are usually centrally collected and combined for discharge at one or a few points.

TYPICAL WASTEWATER SOURCES

Untreated process wastewater

Miscellaneous untreated wastewater - equipment washdown, steam jet condensate, cooling water

Decantates or filtrates

Cleaning wastes

Steam stripping wastes

Acid leaching solutions

Spent plating, stripping or cleaning baths

Spent scrubber, absorber, or quench liquid

Off-spec, discarded products or feedstock

Distillation side cuts

Cyclone or centrifuge wastes

Spills, leaks, vessel overflows

METHODS OF WASTEWATER DISPOSAL

Direct discharge to surface waters

Discharge to a publicly owned treatment works

Underground injection

Surface impoundments

Land treatment

Direct Discharge to Surface Waters

Many facilities discharge wastewater directly to nearby bodies of water; this action requires a National Pollutant Discharge Elimination System (NPDES) permit. The permit usually requires monitoring of the wastewater discharge flow and the concentrations of various constituents within the wastewater (usually generalized constituents such as BOD and TSS). Monitoring is usually not required for most of the individual chemicals or compounds. When such monitoring is required, wastewater flow rate and concentration data collected for the NPDES permit can be used to calculate wastewater releases directly.

Discharge to a Publicly Owned Treatment Works (POTW)

Many facilities discharge their wastewater to POTW's. In some cases, a POTW may require pretreatment of wastewater and/or monitoring of the flow rate and the concentration of various constituents. If a POTW requires monitoring of a chemical or compound subject to reporting, releases of that chemical or compound in the wastewater can be calculated by multiplying the reported concentration by the flow rate. Discharge to a POTW is considered a transfer to an offsite location.

Underground Injection

In some situations, wastewater containing hazardous and/or toxic wastes may be injected beneath the earth's surface in a location where it is unlikely to contaminate ground water. Injection operations are usually controlled by permitting procedures that require maintaining records of the volumes and analyses of the wastes injected. From this information, quantities of listed chemicals and/or compounds that are disposed of in this manner can be directly calculated.

Surface Impoundments

A surface impoundment is a natural topographic depression, man-made excavation, or diked area formed primarily of earthen materials (although it may be lined with man-made materials), which is designed to hold an accumulation of liquid wastes or wastes containing free liquids. Examples of surface impoundments are holding, storage, settling, and elevation pits, ponds, and lagoons. If the pit, pond, or lagoon is intended for storage or holding without discharge, it is considered to be a surface impoundment used as a final disposal method. The operation of surface impoundments is usually controlled by RCRA permits, which requires maintaining records of the volume and concentration of hazardous wastes disposed of. This information can be used for direct calculation of the quantity of a listed chemical and/or compound disposed of in this manner. This disposal method is considered a release to land; however, listed chemicals in the impoundment may be released to air by volatilization, collected as sludge and removed, or biodegraded. Any releases from the impoundment should be accounted for in release totals to air, water, land, or offsite disposal.

Land Treatment

Land treatment is a disposal method in which wastewater is applied onto or incorporated into soil. These operations are usually controlled by RCRA permits with conditions that regulate the volumes of wastewater to be treated, the concentrations of hazardous and/or toxic materials it contains, and the frequency of land application, and also requires a ground-water monitoring program. This information can be used to calculate the quantity of a listed chemical and/or compound disposed of in this manner. Chemicals and/or compounds in the wastewater that are released to the soil or to air (by volatilization) are considered releases to land.

Calculating Releases in Wastewater

Quantities of listed chemicals and/or compounds released to the environment in wastewater can be calculated by summing the releases from individual operations or by determining releases from a central wastewater discharge point (if available). The latter method is preferred because it involves the direct measurement or estimation of the flow of the discharge stream, and the concentrations of chemicals and/or compounds it contains. The following subsections describe the use of direct measurement, mass balance, release data from other facilities in the industry, and engineering calculations to estimate releases of listed chemicals and/or compounds in wastewater. No general compilation of emission factors is available for release in wastewater as it is for releases to air; however, in some instances, information from other facilities in the industry can be applied to estimate releases in wastewater.

Direct Measurement

Direct measurement can be used to calculate releases in wastewater, individual processes or from a central discharge point. This method is used by multiplying the wastewater flow rate by the concentration of the chemical or compound of concern. The following two items describe direct measurement of wastewater releases based on average measured values and multiple measured values, respectively.

Releases Based on Total Annual Volume and Average Measured Concentration

If a wastewater stream has a relatively constant daily flow rate and the measured concentrations of listed the chemicals and/or compounds in the stream do not vary greatly or are well characterized, average values for flow rate and concentration can be used to calculate releases.

Example - Use of Direct Measurement to Estimate Toxic Wastewater Emissions:

Step 1. Gather process information and monitoring data.

A stream containing an average acetaldehyde concentration of 500 milligrams per liter is sent to an onsite treatment system at a rate of 5 gal/min. The stream leaving the treatment system at 5 gal/min contains 25 milligrams of acetaldehyde per liter. If the plant operates 24 hours per day, 330 days per year, the quantity of acetaldehyde entering and leaving the treatment system can be calculated, assuming no net loss of water or acetaldehyde by evaporation to air. Also, the treatment system efficiency can be calculated.

Step 2. Calculate the quantity of acetalde-hyde entering and leaving the system.

Volume = $\underline{\text{5 gal}}$ x $\underline{\text{60 min}}$ x $\underline{\text{24 h}}$ x $\underline{\text{330 days}}$ = 2.38 millon gal
           min   hour   day   year   year

Into system:

$\underline{\text{2.38 mil gal}}$ x $\underline{\text{500 mg}}$ x $\underline{\text{3.78 liters}}$ x $\underline{\text{1 lb}}$ = $\underline{\text{9330 lb}}$
        year   liter   gallon   453,000 mg   year

From system:

$\underline{\text{2.38 mil gal}}$ x $\underline{\text{25 mg}}$ x $\underline{\text{3.78 liters}}$ x $\underline{\text{1 lb}}$ = $\underline{\text{496 lb}}$
        year   liter   gallon   453,000 mg   year

Step 3. Calculate treatment system efficiency.

Treatment system efficiency: $\underline{\text{9930 - 496}}$ x 100 = 95%
                                      9930

Example - Use of Direct Measurement to Estimate Toxic Wastewater Emissions:

Step 1. Gather wastewater flow and con-centration data from the NPDES permit.

The NPDES Permit of a leather tanning facility requires daily monitoring of wastewater flow volume and biweekly analy-sis of a daily composite sample of this discharge for total chromium. The total chromium analytical results for the year are presented below.

Step 2. Calculate releases for those days in which a chromium analysis was performed.

The total chromium releases (in pounds per day) to water for a given day at this facility are calculated by mutliplying the daily flow (in million gallons per day) by the total chromium concentration (in micrograms per liter) times a conversion factor (8.34×10^{-3})

Discharge flow rate 10^6 gal/day	Total chromium, μg/liter	Releases, lb/day
0.415	918	3.2
0.394	700	2.3
0.417	815	2.8
0.440	683	2.5
0.364	787	2.4
0.340	840	2.4
0.457	865	3.3
0.424	643	2.3
0.463	958	3.7
0.414	681	2.4
0.476	680	2.7
0.431	627	2.3
0.369	807	2.5
0.392	729	2.4
0.323	964	2.6
0.302	722	1.8
0.358	566	1.7
0.322	510	1.4
0.330	630	1.7
0.322	630	1.7
0.408	652	2.2
0.442	649	2.4

Discharge flow rate 10^6 gal/day	Total chromium, μg/liter	Releases, lb/day
0.442	649	2.4
0.356	695	2.1
0.390	758	2.5
0.423	658	2.3
0.487	970	3.9
	Average	2.44

Step 3. Calculate annual releases.

Based on an average daily release of 2.44 lb over the year and 250 days of discharge during the year, the yearly total chromium discharged water is:

$$\underline{2.44 \text{ lb}} \times \underline{250 \text{ days}} = 610 \text{ lb per year}$$
$$\text{day} \qquad \text{year}$$

Example - Use of Direct Measurement to Estimate Toxic Wastewater Emissions:

Step 1. Gather analytical results and determine average value.

The results of 10 copper analyses are expressed in micrograms per liter:

6	<5	<5	<5	<5
10	<5	<5	<5	<5

The average concentration is:

$$\frac{1(6) + (10) + 1(8) + 7(5/2)}{10} = \frac{4.2 \text{ micrograms}}{\text{liter}}$$

$$= 4.2 \times 10^{-6} \text{ grams per liter}$$

Step 2. Determine annual releases.

For an annual flow of 37.8 million liters (million gallons), the average discharge would be 4.2×10^{-6} liters/year $= 159$ grams/year or 0.35 lb/year.

CHAPTER 10

REGULATORY COMPLIANCE: AN OVERVIEW
OF WORKER PROTECTION AND RIGHT-TO-KNOW

INTRODUCTION AND OVERVIEW

Environmental compliance is a very complex process--an uncomfortable blend of science and law, engineering and economics. It also demands familiarity with federal, state, and local requirements. The rules are complicated, frequently overlapping, and rapidly changing. Firms large and small encounter grave difficulty in determining what to do: goals, objectives, techniques, priorities.

In the face of this confusion, how do you allocate your available resources of people, money, and technology? Often it's to respond to the latest inspection or notice from whichever regulator has targeted you. If your facility is caught in this cycle, attempts to comply are inevitably reactive and sporadic, and frequently too late. The only way out of this trap is to find time, somehow, to familiarize your organization with all the requirements the company faces. Chapters 10, 11 and 12 are designed to familiarize the Professional Hazards Manager with the environmental and safety regulations. First, the manager must be aware that there are 12 major arenas of environmental management. This chapter starts with worker safety and Right-To-Know. Chapter 11 covers the critical **federal** compliance requirements involving

hazardous materials and hazardous wastes. Finally, Chapter 12 provides an overview of the major environmental statutes. All three chapters highlight the **state** laws modeled on federal requirements and the state agencies that implement state (and often federal) provisions.

WORKER PROTECTION AND OSHA

Most workplaces in the U.S. are regulated under the provisions of the federal **Occupational Safety and Health Act of 1970 (the Federal OSH Act)** and its many state counterparts. The Federal OSH Act assigns national worker protection authority to the U.S. **Occupational Safety and Health Administration (OSHA)**, a part of the U.S. Department of Labor. The Federal OSH Act allows OSHA to delegate its authority to states which have developed a "state plan" for protecting workers with provisions at least as stringent as those provided by OSHA. Half the states in the U.S. have approved state plans in which OSHA retains only oversight authority; OSHA implements federal provisions directly in the other states.

Most of OSHA's worker protection requirements are drawn up into "**OSHA Standards**" that deal with individual topics. Some were developed by OSHA, but many adopt and incorporate numerous models and standards authored by other professional organizations such as the National Fire Protection Association (NFPA), the National Institute of Occupational Safety and Health (NIOSH), and the American National Standards Institute (ANSI).

Areas the Professional Hazards Manager should be familiar with are the

- Hazard Communication Standard (HCS)

- Hazardous Waste Operations and Emergency Response (HAZ-WOPER) Standard

- Fire Protection Standard

- Chemical Process Safety Management Standard

- Laboratory Standard

In addition, OSHA adopts a number of other standards that cover additional mechanical, electrical, chemical, and operational safety provisions.

The Federal OSH Act provides for both federal and state authority over workplace health and safety. OSHA has initial authority for promulgating national standards, and for administering worker safety provisions nationwide. However, the Federal OSH Act encourages states to assume full responsibility for enforcing their own safety and health laws. States may do so by obtaining federal approval of a state plan which demonstrates that state standards are "at least as effective" as corresponding federal standards, and are being adequately enforced by state authorities. Twenty-one

states (plus **Puerto Rico** and the **U.S. Virgin Islands**) have in this way assumed authority over workplace health and safety.

OSHA does not enforce workplace health and safety directly in these State Plan states, but monitors state actions to ensure that the state's programs continue to meet federal requirements. In particular, these states must adopt comparable standards within six months of federal promulgation of a new or revised standard, and enforce these requirements. In State Plan states, the state programs must be "at least as effective" as, and are often identical to, the federal standard. Individual states with approved plans may, however, enforce stricter or additional workplace standards.

Hazard Communication Standard (HCS)

The **HCS** is designed to ensure evaluation of the hazards of all chemicals present in the workplace, and ensure that both employers and employees receive relevant information about those hazards. The federal HCS preempts incompatible state or local regulations, although many states operate approved state programs. States can and do expand on HCS requirements.

Evaluation and communication of hazards must include at least:

- preparation of Material Safety Data Sheets (MSDSs) by manufacturers

- distribution of MSDSs to users of the hazardous substances

- availability of MSDSs to employees

- labeling of chemical containers in the workplace

- development and maintenance of a written hazard communication program for the workplace

- development and implementation of programs to train employees about hazardous chemicals

The HCS defines a **hazardous chemical** as any chemical which is a health hazard or a physical hazard. **Health hazard** means a chemical for which there is statistically significant evidence based on at least one study conducted in accordance with established scientific principles that acute or chronic health effects may occur in exposed employees. Chemicals that pose a health hazard include carcinogens, reproductive toxins, hepatotoxins, and neurotoxins. Chemicals that pose a **physical hazard** include combustible liquids, compressed gases, organic peroxides, and oxidizers.

The HCS generally applies to any hazardous substance to which employees may be exposed under normal use conditions or in a foreseeable emergency. Some exceptions do apply, however, to products which are already labeled to comply with other laws, or which are packaged and handled in ways that prevent employee exposures.

Operations where employees handle only **sealed containers** of chemicals need only comply with the labeling standards and limited MSDS and employee information requirements.

Several classes of chemicals are **exempt from MSDS and labeling** requirements, in deference to other federal programs that provide similar protection. These are:

- any pesticide registered under the Federal Insecticide, Fungicide, and Rodenticide Act (FIFRA)

- any food, food additive, drug, cosmetic, medical or veterinary device regulated under the federal Food, Drug and Cosmetic Act

- any alcoholic beverage intended for consumption

Additionally, many chemicals **are exempt from HCS**, including:

- hazardous wastes

- tobacco or tobacco products

- wood or wood products

- food, drugs or cosmetics intended for personal consumption by employees while in the workplace

- drugs in their final form for administration to patients

- consumer products packaged for distribution to and used by the general public, provided that employee exposure to the product is not significantly greater than the consumer exposure during normal use.

- articles (defined as manufactured items which have a specific shape or design, and do not result in exposure to hazardous chemicals under normal conditions of use)

MSDSs are the most basic source of hazardous chemical information. The HCS requires chemical manufacturers and importers to develop or obtain an MSDS for each hazardous chemical they produce or import.

Each MSDS must provide at least the following information:

- chemical identity, as used on the container label

- chemical and common names (for a single ingredient)

- chemical and common names of all hazardous ingredients (for mixtures)

- physical and chemical characteristics of the hazardous ingredients (e.g., flash point, vapor pressure)

- physical hazards of the hazardous ingredient (e.g., flammability, reactivity, and potential for explosion)

- health hazards, including exposure symptoms

- primary route(s) of exposure

- Permissible Exposure Limits (PELs), Threshold Limit Value (TLV), or other exposure limits

- whether or not any hazardous ingredient is listed as a carcinogen or potential carcinogen

- any generally applicable safe handling and use precautions, control measures, and emergency and first aid procedures

- name, address, and telephone number of the person who prepares the MSDS, and the date of its most recent version

While the contents of an MSDS are mandated by the OSHA HCS, an approved state-plan state may require additional features. For example, in **California**, MSDSs must comply with the labeling and warning requirements for carcinogens and reproductive toxins under Proposition 65. California also requires MSDSs to describe any potential combustion products of a chemical and their hazards.

Manufacturers, importers, and distributors of chemicals must supply an MSDS with the first shipment of any hazardous chemical to another party, and then supply a new sheet whenever that MSDS is updated. Also, manufacturers must update their MSDSs whenever they obtain new hazard information or change product formulations.

The HCS contains provisions to protect manufacturers from having to disclose trade secrets when meeting hazard communication requirements. The chemical identity of a chemical or constituent may be withheld from an MSDS on the following conditions:

- its deletion is noted

- its health effects and physical properties are included

- the specific chemical identity is made available to health professionals, employees or designated representatives in the event of an emergency.

This information must be disclosed during non-emergency situations to these persons, upon written request, when the information is needed for proper treatment of a medical condition, or to assess the need for special handling procedures. The manufacturer may require a confidentiality statement before releasing the information, or can obtain one after the fact if the disclosure request is an emergency.

Chemical manufacturers, importers, and distributors must ensure that each container of hazardous chemicals (while onsite in and when leaving their workplaces) is labeled with the following information:

- identity of the hazardous substance(s)

- appropriate hazard warnings

- name and address of the manufacturer, importer or other responsible party

The HCS requires **manufacturers and importers** of hazardous chemicals to assess the physical and health hazards of those substances. A manufacturer or importer may rely on basic knowledge of the material, or on existing scientific data about the material, or may conduct a hazard assessment through chemical and toxicity testing. The HCS includes criteria for determining whether or not a chemical is hazardous.

The HCS requires a manufacturer or importer to assess the hazard of products that are chemical **mixtures**. This hazard determination may be made by bulk analysis of the mixture, or by considering the hazardous nature of any of the individual chemicals in the mixture. If the manufacturer or importer chooses to look at individual chemicals, then certain concentration limits apply. The analysis must look at all chemicals that:

- make up 1% or more of the mixture or product, or

- are designated as carcinogenic by OSHA, and make up at least 0.1% of the mixture or product, or

- even if less than 1% of the mixture (or less than 0.1% if carcinogenic) could be released in concentrations that would present a health hazard to workers [for example, exceeds the PEL, Short-Term Exposure Limit (STEL), TLV or other measure of health risks of exposures in the workplace].

Employers must provide information to your employees regarding hazardous chemicals present in the workplace to which the employees may be exposed under normal conditions, or in a reasonably foreseeable emergency. The HCS requires employers to communicate all relevant information to employees via MSDSs, labels and other forms of warning, a hazard communication program, and training.

The required basic source of information on hazardous chemicals are MSDSs. Companies must maintain a file of MSDSs for **all** products that have them. They must provide information to employees and contractors about any hazardous chemicals to which they may be exposed in the workplace. Distributors of hazardous substances must relay hazard information to their customers, contractors, and exposed employees.

Employers must make MSDSs readily accessible to employees in the workplace. Employees or their labor representatives must be permitted to make copies of any MSDS information, although the employer does not have to pay for these copies.

When you or your employees discover that an MSDS is missing, you should request a new MSDS from the chemical product's manufacturer. OSHA requires employers to obtain a new MSDS from the manufacturer "as soon as possible." Some states enforce stricter standards. For example, in **California**, both the employer and the distributor or manufacturer are given deadlines for responding to MSDS requests. Failure to provide an MSDS in a timely fashion is to be reported to the California Division of Occupational Safety and Health (Cal/OSHA).

Employers must make sure that each container of hazardous substances in their workplace is labeled, tagged or marked with the identity of the hazardous substance(s) in the container, as well as appropriate hazard warnings. (Containers which temporarily hold hazardous substances for immediate use during transfers need not be labeled, nor do containers used by a worker for temporary storage of chemicals during a single work shift require labels.) Employers may use signs, placards, process sheets, process sheets, batch tickets, operating procedures or other similar materials for stationary process containers where convenient. The information must be readily available to employees during their work shift. All labels must be written in English, although other languages may be added where advisable.

Employers must develop, implement, and maintain at each workplace a written hazard communication program. The program must include procedures to assure that requirements for labels and other warnings, MSDSs, and employee information and training are all met. It must also include:

- a list of the hazardous chemicals in the workplace, with each chemical referenced to its MSDS

- methods to inform employees about the hazards of nonroutine tasks.

- hazards associated with chemicals located in unlabeled pipes in work areas

The written hazard communication program must be made available on request to employees, their representatives or OSHA officials.

Whenever hazardous chemicals are present in a workplace where employees of other employers (such as subcontractors) may be exposed, the employer responsible for the chemicals must also include in its written hazard communication program procedures to inform and warn these other employers and their employees of the hazards present.

OSHA requires employers to provide information and training to all employees who will work with or be exposed to hazardous substances. Training must be provided at the time of their first work assignment, and again whenever a new hazard is introduced into their workplace. Employees must be **informed** of:

- the requirements of the HCS

- any operations in their work area involving the presence of hazardous substances

- location and availability of the written hazard communication program, including chemical lists and MSDSs

- their right and that of their physician and union agent to receive infor--mation regarding hazardous substances to which they may be exposed

- their right to exercise any of the rights designated to them in the Federal OSH Act without being discharged or discriminated against

Employee **training** must include at least:

- methods to detect the release of hazardous substances into the work area

- physical and health hazards of substances in the work area

- measures employees could take to protect themselves from exposures including appropriate work practices, emergency procedures, and the proper use of personal protective equipment

- explanations of MSDSs, labels, similar documents, and methods of obtaining and using appropriate hazard information

Laboratories are subject to a truncated list of HCS requirements. Laboratories that generate new chemicals need not produce MSDSs for those chemicals so long as they are not intended for commercial use. Labeling standards for laboratory containers are also less stringent. Additional requirements covering laboratory safety are provided under OSHA's Chemical Hygiene Plan (CHP) requirement.

Although OSHA regulations exempt pesticides from the OSHA HCS, FIFRA requires that workers using pesticides receive basic hazard information about those chemicals. The U.S. Environmental Protection Agency (EPA) is working to bring FIFRA hazard communication requirements more into line with the OSHA HCS. Some states already enforce more stringent hazard communication requirements for pesticide users. For example, **California** pesticide regulations require strict hazard communication procedures for pesticide applicators, while **Texas** provides overall hazard communication for agricultural workers.

Emergency Response Planning

OSHA and state worker protection agencies require employers to plan for emergencies in the workplace, and to train employees to perform whatever emergency response roles they are assigned in employers' plans. These requirements are continuing to evolve. In many states, such as **California** and **New Jersey**, these requirements overlap with those covering hazardous materials handling.

As an employer, you are required to implement an **Emergency Action Plan** for emergencies in the workplace. Alternatively, you may use comparable plans developed to comply with other laws. These planning requirements include emergency response training for employees, depending on the extent to which employees are involved in emergency response. Even if you do not permit your employees to assist in handling emergencies and you evacuate your employees from the danger area when as emergency occurs, you are required to comply with and implement an Emergency Action Plan. The Emergency Action Plan must be communicated in writing unless you have 10 or fewer employees in which case you are permitted to communicate the plan to your employees verbally. The Emergency Action Plan should include at least the following elements:

- emergency escape procedures and emergency escape route assignments

- procedures to be followed by employees who remain to operate critical plant operations before evacuating

- rescue and medical duties for designated employees

- the preferred means of reporting fires and other emergencies

- names and job titles of persons/departments who can be contacted for further information or explanation of duties under the plan

In addition, the employer is required to establish an employee alarm system with a different signal for each type of emergency. Furthermore, the employer must train those employees designated to assist in the safe and orderly emergency evacuation of the danger area. All employees must be informed of their responsibilities under the plan when the plan is developed, whenever

the employees' responsibilities or designated actions under the plan change, and whenever the plan is changed.

If your Emergency Action Plan includes emergency response assignments for any of your employees, then additional planning and training requirements apply. OSHA and EPA have issued worker safety regulations covering **routine hazardous waste management operations and emergency responses involving hazardous materials** (including hazardous wastes). These rules are often referred to as the **HAZWOPER** Standard, using a contraction of "hazardous waste operations and emergency response."

The HAZWOPER Standard covers three groups of employees:

- workers expected to respond to emergencies caused by spills of hazardous substances (including wastes) at facilities not covered by the two categories below

- workers at hazardous waste site cleanups

- workers at permitted hazardous waste treatment, storage, and disposal (TSD) facilities

All training must be delivered by qualified trainers. OSHA is developing accreditation procedures for these training sessions; proposed procedures for training for employees at hazardous waste cleanup sites and TSD facilities (but not emergency response) were issued on January 26, 1990.

You must prepare and implement an Emergency Response Plan if any of your employees are involved in emergency response. OSHA defines **emergency**

response as a response effort by employees from outside the immediate release area or by other designated responders (e.g., facility fire brigade) to an occurrence that results, or is likely to result, in an uncontrolled release of a hazardous substance. Specialized training for hazardous materials emergency response is mandatory for any worker whose responsibilities specifically include response to releases of hazardous materials (including hazardous wastes). The requirements of the Emergency Response Plan apply to any employee who may play a role in any emergency response situation. The Emergency Response Plan must include at least all of the following components:

- pre-emergency planning and co-ordination with outside parties

- an outline of personnel roles, lines of authority, training, and communication

- emergency recognition and prevention procedures

- details of safe distances and places of refuge

- site security and control procedures

- evacuation routes and procedures

- decontamination procedures

- emergency medical treatment and first aid procedures

- emergency alerting and response procedures

- methods for critiquing response and follow-up procedures

- information on the location of personal protective equipment and emergency equipment

HAZWOPER training varies with the degree of direct involvement in emergency response, as follows:

- **first responder level of awareness:** sufficient training or experience to demonstrate general understanding of hazardous materials and their risks, and some ability to identify them in the field; knowledge of first responders' duties, and how to notify appropriate agencies and personnel

- **first responder operations level:** at least eight hours of training on the topics described above, or sufficient experience to demonstrate "aware-ness level" knowledge; plus knowledge of basic control, contain-ment, confinement, and decontami-nation procedures associated with available resources

- **hazardous materials technician:** understanding described above (at least 24 hours of training at first responder operations level); ability to implement employer's Emergency Response Plan, including use of appropriate personal protective equipment, and response procedures and technologies; and basic know-ledge of chemical and toxicological terminology and behavior

- **hazardous materials specialist:** at least 24 hours of training at the hazardous materials technician level, plus advanced understanding of site and area-wide response plans; ability to select and use appropriate response equipment and procedures; and ability to develop a site safety and control plan

- **onsite incident commander:** at least 24 hours of first responder operations level training; ability to implement incident command system and site and area-wide emergency response plans; and knowledge of decontamination procedures

The HAZWOPER Standard applies to workers involved in the following types of cleanup operations:

- cleanups required by a government agency, such as Superfund site work

- corrective actions undertaken under RCRA or the federal Hazardous and Solid Waste Amendments of 1984 (HSWA)

- voluntary cleanups at uncontrolled hazardous waste sites undertaken by responsible parties with official recognition by a government agency

The worker safety provisions mandated under the Superfund Amendments and Reauthorization Act of 1986 (SARA) specify a set of **safety practices** to be implemented at cleanup sites, and also establish requirements for **employee training**.

To comply with safety requirements, you must perform all of the following at your cleanup sites:

- prepared a written safety and health program [This program may incorporate plans prepared under other laws, such as RCRA and

OSHA's written hazard communication program.]

- characterized all hazardous waste sites before entry by workers

- organized site control and monitoring

- arranged for medical surveillance of workers involved with hazardous wastes (Annual physical examinations are the standard, although qualified physicians may require more frequent examinations, or may certify the adequacy of examinations no less frequent than biennial.)

- put in place engineering controls, work practices and, if necessary, personal protective equipment adequate to keep exposures to health hazards below OSHA-established PELs (These include provisions for routine handling of drums and containers, and decontamination efforts.)

- established procedures to evaluate and introduce "effective new technologies and equipment for the protection of employees" involved with hazardous waste site cleanups

The training provided for workers at cleanup sites must cover the following topics:

- names of personnel responsible for site safety and health

- onsite safety, health, and hazards

- use of any necessary personal protective equipment

- work practices designed to minimize hazards

- safe use of engineering controls and equipment

- identification of symptoms that might indicate overexposure

- medical surveillance (annual physical examinations are the standard, although qualified physicians may require more frequent examinations, or may certify the adequacy of examinations no less frequent than every two years)

- familiarization with components of the site safety and health plan

The extent of training required varies with the degree of worker involvement with hazardous waste cleanup operations, as follows:

- 40 hours of initial instruction, plus three days of supervised field experience, for all new workers engaged in general hazardous waste site operations and for onsite management and supervisory personnel responsible for any areas and workers covered by these SARA training provisions

- 24 hours of initial instruction, plus one day of supervised onsite experience, for workers who will be involved occasionally in a narrow range of hazardous waste activities, or who work at a site which has been fully characterized and presents minimal risk of uncontrolled exposure

- 8 hours of refresher training annually for every category of worker

If a worker's responsibilities change from the category requiring 24 hours of training to the category requiring 40 hours of training, he or she must receive the additional 16 hours of instruction and two more days of supervised field experience before assuming the new responsibilities. Fully trained employees who change sites need only receive added site-specific training.

General OSHA Standards and Chemicals

OSHA has specific requirements covering various industries and operations. For example, there are OSHA standards for construction, machinery, and laundry operations, among many others.

Other relevant OSHA regulations include standards for personal protective and respiratory devices, safety standards for working with aboveground and underground storage tanks (ASTs and USTs), and methods for safely working with pressurized gases, other gases, and flammable liquids. Fire prevention and firefighting methods are specified for a variety of materials. The regulations provide specific safety standards for different types of workplaces, such as service stations, and bulk receiving and storage facilities. These standards include both safe work practices and engineering controls to reduce the risk of accidents.

As an example, there are general requirements for all machines which cover machine guarding, point of operation guarding, and protection from blades. In addition, there are specific requirements for woodworking machinery, mechanical power presses, mechanical power-transmission apparatus, abrasive wheel machinery, mechanical power presses, mechanical power-transmission apparatus, abrasive wheel machinery, forging machines, and others. OSHA also has specific requirements for welding, cutting, brazing, hand-held equipment, powered platforms, manlifts, vehicle-mounted work platforms, and numerous others. State plans may include more stringent standards. For example, **California** requires permits for most air compressors.

In addition, there are a number of OSHA or authorized state worker protection agencies' standards which apply to the use of particular toxic or hazardous materials. Agency inspectors check to ensure that employers are complying with regulations limiting concentrations of specific airborne contaminants in the workplace, as well as worker safety and health requirements for highly dangerous chemicals.

Substances with specific standards regulating potential health effects include:

- asbestos
- vinyl chloride
- coke oven emissions
- 2-acetylaminofluorene
- 4-aminodiphenyl
- benzidine
- 3,3'-dichlorobenzidine (and its salts)
- 4-dimethylaminoazobenzene
- alpha-naphthylamine
- beta-naphthylamine
- 4-nitrobiphenyl
- beta-propiolactone

- bis-chloromethyl ether
- methyl chloromethyl ether
- ethyleneimine
- 1,2 dibromo-3-chloropropane (DBCP)
- acrylonitrile
- inorganic arsenic
- formaldehyde
- benzene
- ethylene oxide
- N-nitroso dimethylamine
- lead
- cotton dust

Each substance is regulated for health effects under guidelines that may include the following:

- PELs/STEL
- air monitoring requirements
- compliance methods
- respiratory protection methods
- protective clothing and equipment requirements
- housekeeping practices
- hygiene facilities and practices
- medical surveillance requirements
- medical removal protection for overexposed workers

- employee information and training for specific hazards

- sign requirements

- record-keeping requirements

- observation and monitoring rights for employees and their representatives

Substances that have specific standards regulating safety issues include:

- acetylene

- anhydrous ammonia

- ammonium nitrate

- cadmium

- calcium carbide

- hydrogen

- lead

- liquid fuels

- liquified petroleum gases

- n-nitrosodimethylamine

- oxygen

- zinc

Each substance is regulated for safety issues under guidelines that may include the following:

- ventilation requirements

- handling requirements

- storage requirements

- safety device requirements

- location requirements--confined spaces, indoors, outdoors

Fire Protection Standard

The Fire Protection Standard establishes basic fire prevention and fire protection requirements, and provides employers with several options for meeting these requirements. **Fire extinguishers** are the most basic pieces of fire protection equipment. Some OSHA standards require subject employers to prepare **Fire Prevention Plans**.

Employer responsibilities under the Fire Protection Standard depend on whether or not they provide portable fire extinguishers in the workplace and who, if anyone, is permitted to use them.

If your company provides fire extinguishers in the workplace, but they are not intended for use by employees, the employer must comply with all of the following:

- establish and implement an Emergency Action Plan

- establish and implement a Fire Prevention Plan

- inspect, maintain, and test all portable fire extinguishers in the workplace according to OSHA standards (i.e., monthly visual inspections and annual maintenance checks)

- conduct hydrostatic testing of all extinguishers according to OSHA standards

If your company provides fire extinguishers in the workplace for use by only certain employees, the employer must comply with all of the following:

- establish and implement an Emergency Action Plan which designates only certain employees authorized to use the available portable fire extinguishers and requires all other employees in the fire area to immediately evacuate upon sounding the fire alarm

- comply with general OSHA portable fire extinguisher requirements such as: mounting, locating, and identifying extinguishers so they are readily accessible to employees; providing only approved fire extinguishers; and maintaining extinguishers in a fully charged and operable condition at all times

- inspect, maintain, and test all portable fire extinguishers in the workplace according to OSHA standards (i.e., monthly visual inspections and annual maintenance checks)

- conduct hydrostatic testing of all extinguishers according to OSHA standards

- provide training and education on the appropriate use of fire extinguishers upon initial assignment and annually thereafter

When any OSHA standard requires a Fire Prevention Plan, it must be a written plan containing at least the following information:

- a list of all major **workplace fire hazards** and their proper handling and storage procedures, potential ignition sources in the workplace and their control procedures, and type(s)

of fire protection systems that can control a fire involving these hazards

- **names or job titles** of personnel responsible for maintaining fire prevention equipment or systems, or controlling ignitions or fires

- **names or job titles** of personnel responsible for control of fuel source hazards

- **housekeeping** procedures for controlling accumulations of flammable and combustible wastes

- **maintenance** procedures for fire protection equipment and systems

- information about the **training** and roles of all new employees

The Chemical Process Safety Management Standard

The Chemical Process Safety Management Standard (the Standard) is designed to prevent or minimize the consequences of catastrophic accidents involving the release of highly hazardous chemicals. As of May 26, 1992 the Standard applies to all workplaces that process certain highly hazardous chemicals above threshold amounts. These chemicals and their thresholds are contained in a separate list which is not the same as EPA's list of Extremely Hazardous Substances, or the accidental releases planning list under the 1990 Clean Air Act Amendments. There are currently 137 chemicals on this list, with threshold quantities varying between 100 and 15,000 pounds. In addition, the Standard covers the manufacture, storage, use, transportation, and sale of explosives, blasting agents, and pyrotechnics. Under certain handling and storage conditions, the Standard also applies to large quantities of flammable liquids and gases.

Compliance requirements are triggered when the volume of a listed chemical exceeds the threshold within a process at one point in time. A "**process**" is defined as "any group of vessels which are interconnected and separate vessels which are located such that a highly hazardous chemical could be involved in a potential release."

The Standard is quite detailed, and requires covered employers to:

- gather process safety information on the hazards of the chemicals used and the process technology and equipment involved

- perform process hazard analysis to identify, evaluate, and control the hazards. The Standard lists specific issues that must be addressed, provides a schedule for conducting the initial analysis and periodic updating, and requires the employer to respond to the findings from the analysis

- consult with employees. A written plan for involving workers in the analyses and other elements of the Standard is required, and workers must be provided with access to the information developed.

- implement written operating procedures, including clear instructions to workers for safely conducting their activities involving the highly hazardous chemicals

- provide employee training, initially for all currently involved in the process, plus workers when they are newly assigned to it. Employers must provide refresher training every three years, and determine that the workers have received and understood the training.

- inform contract employers of the potential hazards. Contract employers must then ensure that their workers are appropriately trained.

- perform pre-startup safety review for any new process facility and facility modifications that cause changes in the process safety information required by the standard. These changes would include ones that outdate existing process safety information on process chemicals, technology or equipment.

- ensure the mechanical integrity of critical process equipment to ensure that chemicals are not released in an uncontrolled manner. This can be accomplished through process inspection and testing consistent with recognized and generally accepted good engineering practices. Employers must correct identified deficiencies in a safe and timely manner.

- issue "hot work" permits for work, such as welding, that may cause ignition near a covered process. These must document implementation of fire prevention and protection standards, authorized work date, and the specific equipment on which the work is to be performed.

- establish written procedures to manage changes to covered facilities, processes, chemicals, technologies, and equipment

- investigate every incident that could reasonably have resulted in a catastrophic release

- implement emergency planning, including, at a minimum, the requirements of the Emergency Action Plan

- evaluate facility compliance with this Standard at least every three years via an audit by a person knowledgeable in the process. Employers must respond to the audit's findings.

- provide all information, including trade secrets, to employees and persons performing the required analyses and audits

Laboratory Standards

In 1990, OSHA issued final rules establishing occupational exposure and training standards for laboratories. These rules incorporate and expand upon the general HCS requirements. The regulations include provisions requiring the development and implementation of onsite Chemical Hygiene Plans (CHPs) to identify and minimize hazards of exposures to chemicals in the laboratory. Existing facilities were to prepare CHPs by January 31, 1991 and new facilities must have CHPs when they begin operation.

OSHA's Laboratory Standard supersedes other occupational safety standards and applies to "laboratory uses of hazardous chemicals." The definition of "laboratory uses" is quite specific. The Laboratory Standard applies to activities that are:

- "laboratory scale," that is, chemical manipulations on a scale that may be easily handled by one person

- processes using multiple chemicals or multiple chemical procedures

- activities that are not part of a production process or that do not simulate a production process

- performed using laboratory-type protective equipment and methods

Laboratory operations that do not expose employees to hazardous chemicals are not covered by the Laboratory Standard. Such operations include using pre-prepared test strips for biological tests, or the use of commercial prepared test kits (e.g., pregnancy tests) where all reagents are contained in the kit.

If your company operates a laboratory where employees are exposed to hazardous chemicals, the employer must develop and implement a CHP. A CHP must be able to:

- protect employees from health hazards associated with their laboratory activities

- keep chemical exposures below regulated levels

A CHP must include the following components:

- **standard operating procedures (SOPs)** relevant to laboratory safety.

- criteria to determine and implement **control measures** to reduce employee exposures. These measures may include engineering controls and personal protective equipment.

- steps to ensure that laboratory **fume hoods and other protective equipment** are operating properly.

- criteria for defining laboratory operations that require **prior approval** from the employer or a supervisor before an employee may proceed. This requirement is intended to protect unqualified employees from conducting particularly hazardous operations.

- provisions for additional employee protection for work with **particularly hazardous substances** including **select carcinogens, reproductive toxins,** and substances with **high acute toxicity.**

- designated personnel responsible for implementing the safety plan, including the assignment of a **Chemical Hygiene Officer** and, if appropriate, a **Chemical Hygiene Committee.** This requirement assures that the CHP will become a working document to improve workplace safety.

- **employee information and training** protocol to ensure that employees understand laboratory hazards and their roles in laboratory safety.

- provisions for **medical consultations and examinations** for affected workers.

Employee information and training programs must ensure that employees are apprised of the chemical hazards in their work areas. This information must be provided at the time of an employee's **initial assignment** to a work area, and whenever new chemical exposures may take place. The frequency of **refresher training** is left to the employer's discretion.

Employers must provide all laboratory employees who work with hazardous chemicals with the opportunity to receive medical attention, including follow-up examinations whenever an employee develops signs or symptoms associated with a laboratory exposure, or whenever monitoring data indicates an exposure requiring medical examination has occurred. Employers must make available medical consultation whenever an event, such as a spill, leak or explosion takes place that may have exposed employees to hazardous chemicals. This consultation takes place to determine the need for further examination.

All examinations and consultations must be conducted by qualified medical personnel and be provided at no cost to the employees. These services must also be provided without loss of pay to the employees, and at a reasonable time and place.

Employers must ensure that incoming chemical containers are properly labeled and the MSDSs are available for all hazardous chemicals used in the laboratory. For new chemicals developed in the laboratory, the employer must determine whether the new chemical is hazardous and, if so, provide appropriate hazard information to employees. Additionally, employers must provide (at no cost to employees) respirators whenever use of these devices is required to keep employee exposures below PELs. Finally, employers must establish and maintain accurate records of exposure monitoring results, medical examinations, and medical consultations for each employee.

Hazardous Waste Operations

OSHA and state worker protection agencies impose training requirements for employees involved in "routine hazardous waste operations." OSHA uses "**hazardous waste**" as defined in RCRA, while state worker agencies apply their state definition of hazardous waste, which may be broader. These requirements are part of OSHA's HAZWOPER Standard.

Under RCRA, if your facility is a generator of hazardous waste you must provide your employees with classroom or on-the-job training which teaches them to perform their duties in a way that ensures worker safety, appropriate emergency response, and the facility's compliance with hazardous waste management rules. Workers must receive initial training within six months of their hiring and may not work unsupervised prior to completion of the training. Refresher training must be given annually thereafter.

The HAZWOPER Standard applies extensive additional training requirements to workers at hazardous waste TSD facilities regulated under RCRA or state laws in three areas:

- implementation of safety practices

- provision of initial and refresher training covering safe performance of regular job duties by routine hazardous waste handlers

- emergency response training for routine hazardous waste handlers

TSD employers must take the following measures to protect their workers who routinely handle hazardous wastes:

- develop and implement a written safety and health program. This program may incorporate plans prepared under other laws, such as RCRA and OSHA's written hazard communication program

- implement a hazard communication program

- implement a medical surveillance program; annual physical examinations are the standard, although qualified physicians may require more frequent examinations, or may certify the adequacy of examinations no less frequent than every two years

- develop procedures for decontamination, container handling, and introduction of new technologies

- develop an emergency response program, including an Emergency Action Plan, an Emergency Response Plan, and employee training

A TSD facility employer, must provide adequate training in routine handling of hazardous wastes to enable employees to perform their duties in a safe and healthful manner. The employer must ensure that:

- 24 hours of initial training is provided for routine handlers of hazardous wastes

- the training is given by qualified trainers

- employees receive eight hours of refresher training annually thereafter

Emergency response training for routine handlers at TSD facilities must include all of the following components:

- hazard recognition

- methods of minimizing risks

- proper use of control and personal protective equipment

- proper operating procedures at incident scenes

- coordination with other employees

- response to overexposure or injury

- Emergency Response Plan

- standard operating procedures

Respiratory Protection Standards

Employers are required to provide employees with suitable and applicable respiratory equipment when such equipment is necessary to protect the health of the employee. Additionally, the employer is required to establish and maintain a written **Respiratory Protection Program** which should include the following information:

- written standard operating procedures governing the selection and use of respirators

- selection criteria for respirators, based on the hazards to which employees are exposed (Respirators shall provide adequate respiratory protection against the particular hazard for which they are designed in accordance with standards established by competent authorities such as the U.S. Department of Interior, Bureau of Mines, and the U.S. Department of Agriculture)

- training and instruction for all users in the proper use of respirators and their limitations

- schedule for regular cleaning and disinfecting of respirators. Respirators used by more than one worker should be thoroughly cleaned and disinfected after each use.

- convenient, clean, and sanitary location for storage of respirators

- regular inspection and maintenance schedule to keep all respirators in operating condition (Respirators designated for emergency use should be thoroughly inspected at least once a month and after each use.)

- appropriate surveillance of work area conditions and degree of employee exposure or stress

- schedule for regular inspection and evaluation to determine effectiveness of the program

- physical suitability of persons assigned to tasks requiring the use of respirators (The local physician will determine what health and physical conditions are pertinent for persons who are to perform the work and use the equipment. The respirator user's medical status should be reviewed periodically.)

Inspection and Enforcement

OSHA enforces its standards and regulations in states that do not have an OSHA-approved program. OSHA's powers of inspection, enforcement, and citation cover all occupational safety and health regulations. OSHA inspectors are authorized to enter factories, plants, and other work environments to ensure that employers follow OSHA procedures. Officers may take samples and photographs, and may interview employees during inspections. **State worker protection agencies generally follow similar procedures, using similar enforcement powers**.

OSHA may issue citations or notices of **de minimis violations** upon review of inspection reports that document violations of OSHA rules or standards, even if the violation has no direct or immediate relationship to safety or health. A citation must describe the violation for which it is issued, and prescribe a reasonable time to correct the violation. Employers must post the citation at the site of the violation or in a place accessible to employees. A civil penalty of up to $7,000 may be assessed for each serious or nonserious violation for which a citation is issued. Any employer who fails to correct a violation within the time allotted may be subject to a $7,000 fine for each day past that limit. A civil penalty of up to $7,000 may also be charged for each violation of OSHA requirements for posting information. Employers with willful or repeated violations may be assessed civil penalties of not more than $10,000 for each violation. Any person who gives advance warning of an inspection may be subject to a fine of up to $7,000 and up to six months' imprisonment.

EMERGENCY PLANNING AND COMMUNITY RIGHT-TO-KNOW (SARA TITLE III)

Title III of the Superfund Amendments and Reauthorization Act of 1986 (SARA) establishes several different reporting and planning requirements for businesses that handle, store or manufacture certain hazardous materials. These reports and plans provide federal, state, and local emergency planning and response agencies with information about the amounts of chemicals businesses use, routinely release, and spill. They also provide the public and local governments with information about chemical hazards in their communities, as well as providing information necessary to enable emergency responders to respond safely to accidents at these facilities.

Specific requirements include:

- planning for emergency response (SARA Sections 301 to 303)

- reporting chemical inventory (SARA Sections 311 and 312)

- reporting ongoing releases of "toxic chemicals" (SARA Section 313)

- reporting leaks and spills (SARA Section 304)

Title III requires states to develop emergency planning programs. In addition, many states have developed additional emergency planning, release reporting and/or community right-to-know programs, some more elaborate and restrictive than federal requirements.

Title III requires each state to create an emergency response commission (SERC). In many states, agencies created by state law prior to Title III were named as SERCs. For example, the **Pennsylvania** Emergency Management Council, the **Illinois** Emergency Services and Disaster Agency, the **Mississippi** Emergency Management Agency, and **Delaware's** Commission on Hazardous Materials all predate Title III and were designated as SERCs. SERCs have responsibilities under all four of the Title III programs addressed in this section.

Title III requires each SERC to establish local emergency planning districts (LEPDs) within their state and local emergency planning committees (LEPCs) for each district. In your state, existing political subdivisions (e.g., counties) may be designated as LEPDs. LEPCs include representatives from a wide variety of organizations operating within the LEPDs, including: state and local officials, broadcast and print media, healthcare, transportation, community groups, and facility owners and operators.

Each LEPC establishes procedures to handle requests from the public for right-to-know information required by SARA Title III. Title III requires that each LEPC prepare a local emergency response plan for its district. These local plans are designed for emergency response to releases of "extremely hazardous substances" (EHS). LEPCs submit these plans to their SERC for approval.

Businesses are not required to prepare a local emergency response plan. This is the responsibility of the district's LEPC. However, businesses may be contacted to provide information to be included in the plan. These local regional plans are typically updated annually. Each local plan includes:

- identification of facilities subject to Title III reporting requirements

- identification of additional facilities contributing to or subject to additional risk due to their proximity to subject facilities

- transportation routes for EHS

- methods to identify releases

- emergency response procedures for facilities and local governments

- emergency notification procedures

- descriptions of emergency response equipment and personnel

- evacuation plans

- designation of community and facility emergency response coordinators

- training programs for emergency response and medical personnel

- schedules and methods to revise, update, and exercise the plan

An EHS is any chemical on the list prepared by EPA, which includes the threshold planning quantity (TPQ) for each listed chemical. This list has been revised several times and and currently lists 360 chemicals. There are six different TPQs, all based on weight: 1, 10, 100, 500, 1,000, and 10,000 pounds. All TPQs must be reported in pounds, regardless of whether the substance is a solid, liquid, or gas. One must calculate amounts of EHS in pounds.

Calculate the weight of EHS in mixtures individually, based on the percentage of the EHS in the mixture. These calculations are required if the mixture contains more than 1% of an EHS, or more than 0.1% of a carcinogen listed as an EHS. If your company handles or stores any EHS in a quantity that meets or exceeds its TPQ, one must comply with emergency planning requirements. Companies are subject to this requirement if you reach the TPQ at any time. If your company handles any EHS at or in excess of its TPQ, you must notify your SERC and LEPC that you are subject to SARA Title III emergency planning requirements. A company must then select a representative from the facility to act as facility emergency response coordinator. The coordinator participates in the local emergency planning process.

Some states have additional requirements for EHS handlers. In **Arizona**, for example, any facility handling EHS in quantities equal to or greater than the federal TPQs is required by state law to prepare an emergency plan as described in the Arizona Revised Statutes Title 26. The state publishes an *Emergency Response Plan Questionnaire* that, when completed properly, satisfies the requirements for the emergency plan. Another example can be found in **California**, where the state has identified a list of acutely hazardous materials (AHM), which is identical to the federal EHS list. Any facility that handles AHM at or in excess of the federal TPQs for those chemicals must register with its local administering agency (AA), using a form developed by the state. After a business has registered, its local AA

determines whether it must comply with additional requirements to minimize risks involved with handling AHM. If the local AA deems it necessary, the handler may have to develop a facility Risk Management and Prevention Program (RMPP).

For facilities that handle hazardous materials, OSHA requires that material safety data sheets (MSDSs) be available. If a facility is required to have MSDSs, then it may be subject to the SARA Title III requirement for reporting MSDS chemicals stored or used. Companies must prepare and submit to the SERC, LEPC, and local fire department their choice of either the MSDSs for each hazardous chemical present at the facility during the preceding year or a list providing specified information about those chemicals. **This Section 311 requirement covers virtually every chemical at the facility, not just the EHS that trigger emergency response planning requirements.** The facility MUST comply with these requirements if the facility handled in the preceding year:

- 10,000 pounds or more of any hazardous chemical required to have an MSDS, **OR**

- any EHS in quantities equal to or greater than either 500 pounds or the federal TPQ, whichever is less

Note that one must calculate the weight of EHS in mixtures individually, based on the percentage of the EHS in the mixture.

The term **"facility"** includes any building, equipment, structure, and other stationary item which is located on a single site or on adjacent sites owned or operated by the same person. Warehouses are included if they are located on the same site as facilities that meet these definitions. Facility also includes motor vehicles and aircraft.

Research laboratories and users of most household and agricultural products are exempt from SARA Section 311 reporting requirements.

The MSDS chemical list must include the chemical and common names of each hazardous chemical used or stored onsite in quantities equal to or greater than 10,000 pounds, or for any EHS, equal to or greater than 500 pounds or its TPQ, whichever is less, during the previous year. Chemicals are to be grouped according to five hazard categories:

- acute (immediate) health hazard

- reactivity hazard

- chronic (delayed) health hazard

- sudden release of pressure hazard

- fire hazard

For most mixtures, one must indicate any hazardous component of each chemical reported on this list. One should also indicate (if appropriate) whether they elect to withhold trade-secret information. MSDS lists do not include information on the **quantity** of material you used or stored onsite.

Some states do have additional requirements regarding MSDS chemicals. In **New Jersey**, for example, the state Department of Health (DOH) prepares a hazardous substance fact sheet for each "hazardous substance" as defined by DOH. (Until the publication of the most recent regulations, DOH referred to these substances as "workplace hazardous substances.") Each fact sheet must contain information comparable to that found on an MSDS. DOH must transmit relevant fact sheets to employers, who must incorporate them into their worker notification and education programs.

The **Illinois** Toxic Substances Disclosure to Employee Act requires employers to notify local fire departments of the presence of "toxic substances," which include MSDS chemicals, and to submit basic emergency response information. Worker protection laws in **Michigan** require employers handling MSDS chemicals to provide lists of these chemicals (including the locations of specific chemicals) upon request to the Michigan Department of Labor and/or local fire departments. Most **North Carolina** employers who store 55 gallons or 500 pounds or more of any hazardous chemicals requiring MSDSs must provide a list of these chemicals to local fire departments, and to any North Carolina citizen within 10 days after a request. The fire department can then require the facility to prepare an emergency response plan.

Facilities must comply with these SARA emergency and hazardous materials inventory reporting requirements if they handle hazardous materials in quantities equal to or in excess of the following:

- 10,000 pounds of any hazardous materials, OR

- 500 pounds or the federal TPQ for any EHS, whichever is less.

Some states, such as **Montana** and **Utah**, accept the less detailed **Tier One** Emergency and Hazardous Chemical Inventory (Tier One) form for Section 312 reporting. Tier One forms do not require specific information about each chemical, but require information about the aggregate chemicals divided into the five hazard classes discussed earlier.

Most state and local agencies, however, prefer (and many require) that these inventory reports be made on federal **Tier Two** Emergency and Hazardous Chemical Inventory (Tier Two) forms or their equivalents. Tier Two forms include the following information:

- the chemical or common names of each hazardous chemical (as listed on the MSDS submittal) and the Chemical Abstracts Service (CAS) number

- the estimated maximum amount, in weight ranges, of each chemical present at the facility during the preceding year

- the estimated average daily amount, in weight ranges, of each chemical

- a description of the manner of storage of each hazardous chemical (e.g., the type of container, whether it is under pressure, its temperature)

- the location of each chemical

- an indication of whether any of this information is to be withheld from the public as a trade secret

Exemptions exist for:

- any food, food additive, color additive, drug or cosmetic regulated by the U.S. Food and Drug Administration (FDA)

- any hazardous chemical that is in solid form in a manufactured product which under normal conditions of use would not cause the user of the product to be exposed to that chemical

- any hazardous chemical that is used for personal, family or household purposes, or is packaged in the same form as a product distributed to and used by the general public (packaging, not use, triggers this exemption)

- any hazardous chemical used in a research lab, hospital or other medical facility under direct supervision of a technically qualified individual

- any hazardous chemical used in normal agricultural operations or any fertilizer sold by a retailer to the ultimate customer

SARA Section 313 (covering toxic chemical release inventory requirements) applies only to certain facilities handling large quantities of chemicals from a specific set of "toxic chemicals." Facilities are classified according to their SIC code number. The Standard Industrial Classification (SIC) code system is based on the principal activities underway at a business or facility. For Section 313 purposes, you should determine the SIC Code for each facility separately. If you do not know your SIC code, check with your insurance broker, legal counsel, trade association, local chamber of commerce or the state department of labor. If the facility employs fewer than 10 full-time people, it is exempt from Section 313. Note that each facility must be considered separately.

The EPA creates and maintains a list of "toxic chemicals" which are potentially subject to SARA Section 313 reporting requirements. This list is updated regularly. The most recent version of this EPA list is in the 1991 Form R Reporting Package. If you use any chemical on this list above the reporting thresholds, you must file a report. When new chemicals are listed by EPA before December 1 of any year, your releases of those chemicals must be reported for that year in your SARA 313 report due the following calendar year. Releases of new chemicals listed on or after December 1 do not have to be reported until the subsequent calendar year.

Reporting thresholds for each listed toxic chemical are based on two sets of annual volumes:

- 25,000 pounds per year for toxic chemicals you **manufactured** or **processed**

- 10,000 pounds per year for toxic chemicals you otherwise **used** during a calendar year

The term "**manufacture**" means to produce, prepare, import or compound a toxic chemical. "**Process**" means the preparation of a toxic chemical, after its manufacture, for distribution in commerce. "Process" includes making mixtures, repackaging or using a toxic chemical as a feed-stock, raw material or starting material for making another chemical.

Exemptions exist for those toxic chemicals:

- in concentrations of less than 1% of a mixture or less than 0.1% of a mixture when the chemical is an OSHA-defined carcinogen

- that are structural components of your facility

- that are in food, drugs, cosmetics or other items for personal use

- that are used in motor vehicle maintenance

- that exist as solids in manufactured items and do not release toxic chemicals under normal conditions of use

- that are used in process water and noncontact cooling water drawn from the environment or municipal sources

- that are in intake air used either as compressed air or as part of combustion

- that are used in a research lab, hospital or other medical facility under direct supervision of a technically qualified individual

A Form R report (its full name is "Toxic Chemical Release Inventory Reporting Form") asks for estimates of your total releases of toxic chemicals into the air, water, and land. These releases may be routine or accidental.

Form R asks for a broad range of information about the facility, activities, and chemical uses, including the following:

- the facility's name, location, and principal business

- certification by the owner, operator or "senior management official" of the truth and completeness of information and the reasonableness of assumptions

- for each listed toxic chemical (unless you claim its identity to be a trade secret):

    -- whether the chemical is manufactured, processed or otherwise used
    -- the chemical's CAS number, name or category
    -- an estimate of the maximum amount (in weight ranges) of the chemical onsite **at any time** during the preceding calendar year
    -- an estimate of the total amount of each toxic chemical released onsite during the year, including both accidental releases and routine emissions
    -- transfers of each toxic chemical in wastes to offsite locations
    -- onsite waste treatment methods and their efficiency
    -- onsite energy recovery processes
    -- onsite recycling processes
    -- source reduction and recycling activities

The revised Form R for 1992 incorporates for the first time information requirements mandated by the Pollution Prevention Act of 1990. As of July 1, 1992, if a facility is subject to SARA's Section 313 requirements, it has additional reporting and record-keeping obligations. The facility is required to submit information about its in-house waste minimization efforts for the previous calendar year. This information is included on the revised Form R report.

You are required to report all of the following for the current reporting year (1991) and for the preceding reporting year (1990), and to estimate these quantities for the next two reporting years (1992 and 1993):

- the total quantity of toxic chemicals released, including "any spilling, leaking, pumping, pouring, emitting, emptying, discharging, injecting, escaping, leaching, dumping, or disposing into the environment"

- quantity used for energy recovery onsite and offsite

- quantity recycled onsite and offsite

- quantity treated onsite and offsite

Facilities must also report the following for the current reporting year:

- amount of toxic chemicals released as a result of a one-time event (that is, due to remediation, accident, etc.)

- ratio of reporting year's production to previous year's production

- any source reduction practices used

- whether additional optional information on source reduction, recycling, and pollution control activities is included with your report

Some states do require additional information or have more stringent toxics use reduction goals than the federal program. For example, in 1989, **Massachusetts** adopted one of the most aggressive pollution prevention programs in the country, the Massachusetts Toxics Use Reduction Act (TURA). TURA sets a goal for Massachusetts industry to generate 50% less toxic or hazardous byproducts in 1997 than it generated in 1987. TURA reporting requirements apply to a wider range of facilities than the federal requirements, and TURA requires an additional annual report (Form S). TURA also includes provisions for the preparation of a Toxics Use Reduction Plan. The first Plan is to be submitted in 1994 and must be updated and recertified every two years thereafter.

EPA requires suppliers of listed chemicals, mixtures or most products that contain these chemicals to notify their customers in writing that the chemicals, mixtures or products are subject to Section 313 reporting requirements. Suppliers must include in this notification the name and CAS number (if available) of each chemical purchased by the customer and the percent by weight of each toxic chemical in the mixture or product. If you are a supplier, you must send notifications to each customer with the first shipment in a calendar year.

If you are the owner of a facility covered under SARA Title III, and your only interest in the facility is the real estate the facility sits upon, you are NOT subject to SARA Title III reporting requirements. Instead, the **operator** of the facility is subject to reporting requirements (if the other reporting criteria are met).

A facility is subject to SARA's spill reporting requirements if it releases:

- an EHS that meets or exceeds the EPA-established reportable quantity (RQ--many are as low as one pound)

- a CERCLA-defined hazardous substance above the EPA-defined RQ

A release is any "spilling, leaking, pumping, pouring, emitting, emptying, discharging, injecting, escaping, leaching, dumping or disposing into the environment (including the abandonment or discarding of barrels, containers and other closed receptacles) of any hazardous chemical, extremely hazardous substance or toxic chemical."

Under the federal law, you do **NOT** have to report a release if it:

- is below RQs

- results in exposures only to people onsite at your facility (even if the leak or spill **DOES** exceed RQs)

- is allowed under a federal permit [such as emissions allowed under the Clean Air Act (CAA) or discharges allowed under the Clean Water Act (CWA)], or

- is a licensed use of a pesticide

A release notice must include the following information (to the extent you know it at the time and as long as determining any of the following does not result in delay of notice):

- chemical name or identity of all substances involved in the release

- whether the substance is listed as an EHS

- quantity released

- media into which the release occurred

- known or anticipated acute or chronic health risks associated with the emergency and medical advice, if appropriate

- proper precautions to take in response to the release, including evacuation, if necessary

- name and telephone number of a person to be contacted for further information

As soon as practicable after a reportable release, one must submit a written follow-up emergency notice to the agencies to whom you originally reported. The report should update the information provided at the time of the release and include the following additional information:

- actions taken to respond to and contain the release

- any known or anticipated acute or chronic health effects of the release

- advice, if appropriate, regarding the medical attention necessary for individuals exposed to the release.

Additional follow-up notices must be submitted as more information about the release becomes available.

CHAPTER 11

REGULATION OF HAZARDOUS WASTES

INTRODUCTION AND OVERVIEW

National and state laws provide overlapping structures for what commonly is referred to as "cradle to grave" management of "hazardous wastes." Many states copy federal requirements, and many states have been formally delegated control over federal provisions. Some states have adopted additional provisions (defining more wastes as "hazardous," or exerting more controls over generators, for example) which means that people who manage hazardous wastes must be aware of both sets of requirements applicable to their operations.

The federal program is called "RCRA," the acronym for Resource Conservation and Recovery Act of 1976, even though federal requirements have been enacted by a number of different congressional acts over nearly 30 years. The U.S. Environmental Protection Agency (EPA) promulgates RCRA regulations, and either implements RCRA provisions or delegates them to authorized state agencies. **Georgia** is the only state that operates the entire hazardous waste system. To qualify for delegation, state laws and regulations and their implementation and enforcement must be "at least as stringent" as their federal counterparts. EPA frequently defers to the state on most day-to-day matters, while retaining official independence and preeminence. **As a result, most hazardous waste regulatory**

compliance requirements are frequently dominated by state rather than federal law. For example, the **New Jersey** Department of Environmental Protection (DEP) regulates most aspects of hazardous waste control in lieu of EPA control, while in **California** the Department of Toxic Substances Control (DTSC) administers most aspects of RCRA under interim authorization from EPA. In some states, counties or cities also play roles in the regulation of hazardous waste in concert with state agencies.

Individual state legal provisions and regulations are being amended over time to conform more closely with RCRA changing provisions, so that state agencies can apply for or maintain formal delegation of RCRA authority.

DEFINING HAZARDOUS WASTE

RCRA provides a complicated, overlapping set of definitions of when a material becomes a "waste," and when a waste is "hazardous." States generally copy RCRA's approaches, although some states apply additional or stricter standards. The following series of explanations outlines these complicated approaches.

RCRA defines **waste** as material that is abandoned, recycled, or "inherently wastelike." It could be any discarded material resulting from industrial, commercial, agricultural or community activity. However, pretreated solid or dissolved material in domestic sewage and permitted point source industrial discharges into waterways are covered under control programs developed under Clean Water Act (CWA) legislation and state water quality acts, and are generally exempt from most aspects of hazardous waste regulation.

The statutory definition of **hazardous waste** is a "solid" waste or combination of:

> solid wastes which, because of its quantity, concentration, or physical, chemical, or infectious characteristics, may cause or significantly contribute to an increase in mortality or an increase in serious irreversible, or incapacitating reversible illness; or pose a substantial present or potential hazard to human health or the environment when improperly treated, stored, transported, or disposed of, or otherwise managed.

RCRA defines all hazardous wastes as "solid waste"; this includes all types of hazardous wastes, whether they are solid, semi-solid, liquid or even gaseous (so long as they are in containers). This counter-intuitive definition is a holdover from the first hazardous waste laws which have never been changed.

RCRA provides three basic ways to define a waste as **hazardous**:

- **listed wastes**--wastes included on a specific list of hazardous wastes prepared by EPA

- **characteristic wastes**--wastes which demonstrate hazardous characteris-

tics, considering both physical hazards and adverse health effects.

- **statutory wastes**--those defined under the statute, reflecting certain exemptions and exclusions like the ones listed above

Most states have adopted so-called RCRA tests, and some use additional tests or stricter methods--meaning each state may regulate more hazardous wastes than RCRA prescribes. For example, both **California** and **Washington** provide that all wastes defined as hazardous under RCRA also are state hazardous wastes, but they then add certain tests and stricter standards which cause additional wastes to qualify as hazardous.

RULES FOR GENERATORS

RCRA provides a list of wastes that are defined as hazardous. The EPA has listed approximately 500 specific hazardous wastes that it considers hazardous to human health or the environment. The list is separated into three categories of chemicals:

- hazardous wastes from nonspecific sources ("F wastes")

- hazardous wastes from specific sources ("K wastes")

- discarded commercial chemical products, off-specification species, container residues, and spill residues ("P wastes" and "U wastes")

If any of the wastes generated, including mixtures, contain wastes appearing on these EPA lists, RCRA considers the facility a generator of hazardous wastes.

EPA has established procedures under which the generator of a waste listed as hazardous -- such as the residue from treating or incinerating hazardous materials -- can petition to have these specific wastes delisted. A petition must include analyses of the specific waste, demonstrating two principles to EPA's satisfaction:

- none of the four general criteria apply to the specific waste

- the waste is otherwise safe handled as nonhazardous (for example, no hazardous constituents would leach out of the waste after landfill disposal)

This information typically includes:

- information about the generator

- detailed information about the waste, including its physical description, quantities produced, the generating process, and present waste handling techniques

- information on waste sampling techniques and procedures

- information about the testing laboratory used and the techniques it employs

- laboratory results from at least four representative samples following applicable testing procedures.

Because the delisting process can be quite expensive and time-consuming, it generally is cost-effective only for large wastestreams.

Some state hazardous waste laws provide lists of hazardous wastes in addition to those regulated under RCRA. These extra listings may result because state tests are more stringent, because the state defines additional hazardous characteristics, or simply because there is a state list of hazardous wastes which contains additional wastes. (All of these cases are true for both **California** and **Washington**, for example.)

Many state regulations allow waste generators to determine that their state-defined wastes are not hazardous. If the facility makes such a determination about a **non-RCRA** waste they may contact the state hazardous waste agency to learn the proper procedure to obtain departmental concurrence in that determination. Typically, a generator who wishes to remove their wastes from state regulations must provide information to the appropriate agency about the waste supporting nonhazardous status. Requirements typically are similar to those for RCRA delisting. **Again, because the delisting process can be quite expensive and time-consuming, it generally is cost-effective only for large wastestreams.**

RCRA and state laws also provide that a waste can be hazardous if it exhibits one or more characteristics which make it physically or biologically hazardous. Under RCRA, EPA has defined four such characteristics:

- corrosivity

- ignitability

- reactivity

- toxicity

Some states may add additional characteristics. For example, **California** regulates "persistent and bioaccumulative" chemicals as a subset of toxic wastes, while **Washington** specifically regulates wastes exhibiting "carcinogenicity," "persistence," and a more broadly defined "toxicity."

Generators must analyze nonlisted wastes to determine if they meet any of these hazardous characteristics, and thus qualify as characteristic hazardous wastes. EPA has defined methodologies for testing for these characteristics, and provided for state certification of laboratories qualified to perform one or more of the tests. To ensure that an acceptable methodology is used, only a state-certified laboratory should be contracted for any analysis required. Also note that some states use additional analytical procedures for waste determination.

Characteristics of Wastes

Corrosivity -- A waste is **corrosive** if it dissolves metals and other materials, or burns the skin or eyes on contact. Both EPA's RCRA regulations and state regulations include in this category wastes having either of the following characteristics:

- aqueous wastes with a pH below 2 or above 12.5

- liquid wastes which corrode steel or aluminum at a rate greater than 1.25 inches (6.35 millimeters) per year (using a standard testing protocol)

Alkaline degreasers, spent metal treating and plating solutions, corrosive cleaning solutions, rust removers, waste acids, and bleach compounds (peroxides and chlorine compounds) are examples of corrosive wastes.

Ignitability -- In general, a waste is **ignitable** if it can easily be lighted or, if ignited, burns so vigorously that it creates a hazard. Both EPA's RCRA regulations and state regulations include in this category:

- liquids with flash point below 140°F (60°C)

- nonliquids capable under standard temperature and pressure of causing a fire by means of friction, absorption of moisture, or spontaneous chemical changes and which, when ignited, burn so vigorously and persistently that they create a hazard

- flammable compressed gases

- oxidizers

Paint wastes, nonhalogenated degreasers, thinners and solvents (petroleum distillates), stripping agents, epoxy resins, adhesives, rubber cements and glues, and some waste inks are all examples of ignitable wastes.

Reactivity -- A waste is **reactive** if it is extremely unstable and undergoes rapid or violent chemical reactions such as catching fire, exploding, or giving off fumes when exposed to water or air. This category includes any waste which:

- is normally unstable and readily undergoes violent change without detonating

- reacts violently with water

- forms a potentially explosive mixture with water

- when mixed with water, generates toxic gases, vapors or fumes in a quantity sufficient to present a danger to human health or the environment

- is capable of detonation, explosive reaction or explosive decomposition

- is capable of detonation or explosive reaction if subjected to a strong initiating source or if heated under confinement

Examples of reactive wastes are cyanide compounds from electroplating and metal treating and ore leaching processes, permanganate and manganese wastes from dry cell battery, paint, ink and dye manufacturing, bleaches and hypochlorites from water treatment processes and swimming pools, and other sanitizing operations.

Toxic -- A waste is **toxic** if it contains high concentrations of poisonous heavy metals or certain organic compounds such that it can cause illness or death if inhaled, swallowed or absorbed through the skin. Long-term effects of a toxic waste on human health may include cancer, birth defects, reproductive anomalies, brain and kidney damage, and diseases of the skin, lungs, and heart.

Examples of wastes meeting the toxic criteria are: waste inks and sludges containing certain heavy metals, photographic plating baths and processing wastes, batteries containing lead, certain pesticides, and paint wastes containing heavy metals like chromium or lead.

There has been considerable debate over the years in determining the appropriate test(s) to apply to determine whether a waste is "toxic," and the substances to be tested for.

Traditionally, EPA applied a test for RCRA toxicity called the **"Extraction Procedure--Toxicity Characteristic"** (often referred to as "EP-Tox" or "EPTC") which tested for eight inorganic elements and six organic pesticides.

The number of wastes that EPA deems hazardous due to their toxicity characteristics increased dramatically in late 1990, when EPA replaced the EP Toxicity test with a new **Toxicity Characteristic Leaching Procedure (TCLP)**. The TCLP is a more rigorous procedure than the EP Toxicity test. In addition, EPA's regulations now require testing not only for the 14 toxic chemicals listed under the earlier EP toxicity characteristic, but **25 additional organic contaminants**. These include many common organic chemicals such as benzene, carbon tetrachloride, chloroform, and vinyl chloride. The regulatory levels of these chemicals were determined using health-based criteria and a complex dispersion model to predict their movement in groundwater. The expanded toxicity characteristic initially applied to all large quantity generators (above 1,000 kilograms per month) beginning September 25, 1990. EPA gave small quantity generators (100 to 1,000 kilograms per month) until March 29, 1991 to comply.

States that have been delegated RCRA authority by EPA must use the TCLP for toxicity determinations. However, some states also use additional tests for state-regulated wastes. For example, **California** still uses its own Waste Extraction Test (WET) for a number of non-RCRA hazardous wastes but also provides that TCLP testing be applied to the appropriate wastes. Generators must be aware that the TCLP results in many more wastes being characterized as "toxic" and therefore hazardous.

Additional Characteristics

Some states define additional characteristics for hazardous wastes, for example **persistent or bioaccumulative** toxic wastes. A waste is considered persistent or bioaccumulative if it does not biodegrade or accumulates in a biological organism or system.

Special Categories

A facility may generate certain types of hazardous wastes that RCRA or state laws subject to special waste management requirements because of the particular physical and chemical properties of the wastes, or because of the need to validate existing management practices. These special provisions recognize that not all wastes are equally hazardous. In general, these special provisions increase controls on more hazardous wastes, or relax controls on less hazardous wastes.

Acute Hazardous Wastes

Both EPA and state agencies have defined small groups of hazardous wastes which require additional control measures because the wastes are especially dangerous. EPA defines a class of **acute hazardous wastes (AHW)** which have been found to be fatal to humans in "low doses," have acute oral "LD_{50}" to rats of less than 50 milligrams per kilogram (that is, this is the lethal dose to **50%** of the targets), an inhalation LC_{50} to rats of less than 2 milligrams per liter (the lethal concentration to **50%**), or a dermal LD_{50} to rabbits of less than 200 milligrams per kilogram, or otherwise is considered by EPA to be acutely harmful.

State laws create similar categories for particularly hazardous wastes. Most states retain the AHW name, although some states use different names. [For example, both **California** and **Washington** call AHW "extremely hazardous waste" (EHW); Washington calls other hazardous wastes "Dangerous Wastes."] The health effects standards (such as LD_{50} and LC_{50}) may be easier to achieve under your state's guidelines than under RCRA (this means that wastes which do not qualify as federal AHW may still be subject to stricter standards in a respective state).

RCRA and the state laws provide additional requirements for AHW. Special lower threshold quantities classify one as a small generator under RCRA for AHW. If the facility generates only one kilogram of AHW monthly (or 100 kilogram of any residue or contaminated soil, waste or other debris resulting from the cleanup of a spill into or on land or water containing any acute hazardous substance) full RCRA and state regulatory requirements for hazardous waste generators apply. Some states, including **California** and **Washington** require a special waste disposal permit for AHW or maintain other more restrictive standards.

Special Waste Categories

Some states have the authority to define wastes to be "special wastes." (States that have "Special Waste" categories include **California, Illinois,** and **Washington.**) Though technically hazardous under state standards (though not under RCRA), these wastes contain a high concentration of water (as generated) and/or other materials that dilute the contamination, **or** they are solids that are especially nonreactive in the environment.

Some of the materials that may qualify as special wastes include ash from burning of fossil fuels, auto shredder wastes, catalysts from petroleum refining and chemical plant processes, cement kiln dust, drilling muds from gas and oil wells, sand from sandblasting and foundry casting, slag from coal gasification, mine and ore processing tailings, dewatered sludge from wastewater treatment, and baghouse and scrubber wastes from air pollution control systems.

Waste Oil

Until recently, used oil was not considered hazardous under RCRA. However, using the TCLP, most used oils will exhibit a toxicity characteristic because they typically contain both aromatic hydrocarbons (e.g., benzene) and heavy metals. Because application to used oil of all the requirements applied to most hazardous wastes might severely disrupt ongoing used oil recycling efforts, EPA is developing abbreviated requirements to facilitate continued recycling.

As operated by EPA, used oil that meets one or more toxicity characteristics but is recycled in a manner other than burning for energy recovery or application to the ground for dust suppression is not regulated as a federal hazardous waste; disposal of the same oil without recycling, however, falls under RCRA requirements. EPA anticipates that inclusion of used oil under hazardous waste regulation will actually promote additional oil recycling, since by recycling the oil its generator can avoid the rigorous regulatory procedures.

Used oil may also be "recycled" for energy recovery; regulation of these activities is more stringent than for other recycling activities, but less stringent than the full hazardous waste regulation. Under EPA "hazardous waste fuel" provisions, to qualify for the streamlined requirements, the used oil must have a flashpoint of at least 100°F, contain no more than 1,000 parts per million (ppm) total halogens and no more than five ppm of polychlorinated biphenyls (PCBs).

Meanwhile, many states operate their own programs to facilitate the recycling of used oil. State programs may include provisions for individuals with small quantities of used oil to deliver them to facilities that are not treatment, storage, and disposal (TSD) facilities like service stations for later pick-up. Such systems operate in **Arizona** and **California**. "Milk-run" oil recyclers also operate to retrieve used oil from multiple locations in a single tank truck.

While EPA has struggled to define an appropriate regulatory status and controls for used oil, several states, including **California, Massachusetts,** and **Washington**, define all used oil as hazardous waste. Generally under these programs, used oil that meets certain purity standards does not have to be treated or disposed of in full compliance with standards, but instead may be (sometimes, must be) recycled for use as oil, or burned as a fuel.

Used oil can also be recycled, if it meets purity standards even more rigorous than the standards listed above. Specifically, to be recyclable, RCRA requires that the oil must have no more than the following contaminant levels: lead (50 ppm or less); arsenic (5 ppm or less); chromium (10 ppm or less); cadmium (2 ppm or less); halogens (3,000 ppm or less, if otherwise in compliance with EPA regulations); and the oil must still meet flashpoint standards to its type (e.g., lubricating oil). The concentrations of PCB must be less than thresholds set by EPA (2 ppm) or state agencies. **Oil that meets these standards is not considered hazardous under RCRA unless it meets one of the**

"characteristic" criteria described above.
State laws typically describe a set of procedures for transporting and recycling these oils.

Lead-Acid Batteries

EPA regulations exempt from hazardous waste regulations those who generate, transport, collect or store spent lead acid batteries without reclaiming them. Owners or operators of facilities that store spent lead-acid batteries before reclaiming them, on the other hand, are subject to a reduced set of facility standards.

Many states operate lead-acid battery recycling programs that are more stringent than the federal system. Some states, such as **Illinois** and **California**, require battery retailers to take old batteries in trade for new ones, and to properly recycle old batteries.

Tailings and Mineral Wastes

A 1980 amendment to RCRA (the Bevill Amendment) addressed the regulatory status of wastes from mining and minerals recovery. "Solid wastes from the extraction, beneficiation, and processing of ores and minerals" were excluded from RCRA regulations pending the outcome of several mandated EPA studies.

Through most of the 1980s, EPA interpreted these provisions to cover **all** such wastes, and therefore excluded them from RCRA requirements. In July 1988, EPA changed its approach, deciding that the Bevill Amendment applies only to "special wastes:" defined as those that are both high volume and low hazard. These are generally **not** the same as the "special wastes" discussed above. EPA issued final regula-

tions on September 1, 1989, which defined "high volume" as annual facility generation of 45,000 metric tons of nonliquid wastes or one million metric tons of liquid wastes, and "low hazard" by reference to RCRA-approved leachate tests or pH measurements. These rules retain four mining wastestreams within the Bevill Amendment exclusion, and conditionally retain 20 more. All other mining and mineral extraction wastes now come under RCRA regulation. Many states have adopted these federal requirements.

CLASSES OF HAZARDOUS WASTE GENERATORS

Under RCRA, hazardous waste generators are classified according to **how much hazardous waste they generate during each calendar month**. The three divisions are:

Generators: Those who generate more than 1,000 kilograms per calendar month of hazardous waste. (Referred to as "large quantity generators.") Generators who produce more than 1 kilogram per calendar month of AHW are regulated as large quantity generators as well. Finally, generators who produce more than 100 kilograms per calendar month of soil or other material contaminated with AHW are also regulated as large quantity generators.

Small Quantity Generators (SQGs): Those who generate from 100 up to 1,000 kilograms per calendar month of hazardous waste.

Conditionally Exempt SQGs: Those who generate no more than 100 kilograms per calendar month of hazardous waste, or no more than 1 kilogram per calendar month of AHW or 100 kilograms per calendar month of soil contaminated with AHW.

Large quantity generators are subject to the full range of requirements placed on generators by RCRA. SQGs, while being regulated by RCRA are given a number of less stringent standards to follow. The activities of Conditionally Exempt SQGs are outside the hazardous waste regulation system so long as they follow certain guidelines.

If your facility generates more than 1,000 kilograms (2,200 pounds) of hazardous waste **in any calendar month**, or if it generates more than 1 kilogram (2.2 pounds) of AHW **in any calendar month**, the facility is subject to full RCRA regulation as a hazardous waste generator. As a guide, note that one liter of water weighs one kilogram. There are 3.7854 liters in a gallon, so approximately 265 gallons (about five 55-gallon drums) of water weighs 1,000 kilograms.

If the facility falls into the SQG category, it is granted considerable relief from full generator regulation. State laws, however, may be stricter. For example, **New Jersey and Washington** both regulate generators of more than 100 kilograms per calendar month of hazardous waste as large quantity generators.

The **conditionally exempt** facility falls outside the federal regulation as a hazardous waste generator so long as it determines the hazardous nature of its wastes and takes one of the following five steps when disposing of hazardous waste either onsite or offsite:

- sends the hazardous waste to a facility with a valid TSD facility permit under RCRA or authorized state law

- sends the hazardous waste to a facility with an interim status permit under RCRA or authorized state law

- sends the hazardous waste to a state-permitted, licensed or registered municipal or industrial **solid** waste facility.

- sends the hazardous waste to a facility where it is beneficially used or reused, recycled or reclaimed

- sends the hazardous waste to a facility for treatment prior to beneficial use, reuse, recycling or reclamation

A conditionally exempt SQG may accumulate hazardous waste onsite. However, if the amount of waste accumulated at any one time exceeds 1,000 kilograms, the generator becomes subject to normal SQG requirements for handling that waste. The time limitation for accumulating that waste begins if you exceed this quantity. Remember, though, that your state may enforce stricter regulations. For example, **all** hazardous waste generated in California must be managed through the hazardous waste system; however, a very small generator (less than 100 kilograms per calendar month) may accumulate up to 100 kilograms of waste onsite before other generator standards take effect.

For AHW and soils contaminated with AHW, the generator becomes fully regulated as a large quantity generator after accumulation thresholds are 1 kilogram and 100 kilograms, respectively. If you exceed these limits, you do not qualify as a conditionally exempt SQG. You must fulfill the responsibilities of a hazardous waste generator.

RESPONSIBILITIES OF GENERATORS

A hazardous waste generator must meet **both RCRA and state requirements**. In most states these two sets of requirements are virtually identical.

Each hazardous waste generator must file a "Notification of Regulated Waste Activity" with the state's hazardous waste agency-- EPA or your state hazardous waste agency if it has been granted RCRA authority. (EPA Form 87-1200). This Notification form requires the following information:

- name and address of the generator

- location of generation site

- description of the generation activity

- general description of the hazardous waste being handled

- quantities of hazardous waste handled annually

- name and phone number of a person to contact in the event of an emergency at the site

SQGs are fully subject to this requirement, but conditionally exempt SQGs and households that generate hazardous waste are not.

An EPA Identification Number (ID Number) is used by EPA and by all state hazardous waste agencies to track activities involving hazardous waste. A generator, must obtain an ID Number from EPA or its state agency before transporting, treating, storing or disposing of your hazardous wastes. These numbers are site-specific; this means that a new company taking over a facility should inherit the old ID number (filing an amended notification to alert EPA and/or your state agency to the change), while a moving company would secure a new number, and

a company with multiple facilities would require multiple numbers.

If your only hazardous waste is a state-only hazardous waste (non-RCRA but hazardous under your state laws), you may need to apply for a state ID number from the appropriate state agency even though you don't need an EPA ID number. Most state laws use the EPA ID number as a state ID number as well. Some have additional state numbers. For example, **California** issues its own numbers for non-RCRA hazardous waste generators and also issues Board of Equalization numbers for hazardous waste fee collection purposes.

When shipping wastes offsite, RCRA and state hazardous waste laws require that the generator use only transporters who have ID numbers and that they ship wastes only to permitted TSD facilities (that also have appropriate ID Numbers). A generator can apply for EPA ID numbers using the "Notification of Regulated Waste Activity" form described above.

RCRA and state laws require hazardous waste manifests to assure uniform record keeping for all shipments of hazardous wastes. EPA produces a "uniform hazardous waste manifest." All states use either these forms **or their own state manifests**. (States that use their own manifests include: **Arizona, California, Illinois, New Jersey, New York,** and **Texas**.) For shipments within states with their own manifest, the state manifest must be used; generally when shipping hazardous waste to an out-of-state facility, a generator uses the manifest used **in the receiving state**.

EPA's manifest form requires the generator to provide the following information about each shipment of hazardous waste:

- names, addresses, telephone numbers, and EPA ID Numbers of the generator, transporter, and designated recipient facility (typically a TSD facility)

- descriptions of each hazardous waste in the shipment, including U.S. Department of Transportation (DOT) description, EPA waste code number, type and quantity of all containers, and any additional description (some states, including **California**, also define state waste code numbers which must be included on manifests for shipments originating or ending in that state)

- special handling instructions and additional information, if any

- signed certification by the generator attesting to the truth of the information. Large generators must certify the existence of a waste reduction program, while SQGs certify that they are making good faith efforts to reduce waste generation and are using the best management practices they can afford.

Under RCRA, each shipment requires **at least four copies of the manifest**: enough copies so that the generator, each transporter, and the destination TSD facility each has a copy for its records, plus an additional copy for return from the TSD facility to the generator. (Several states including **California, Illinois, Massachusetts, and Texas** require additional copies.) When shipping hazardous waste, the generator signs and dates each copy. These are presented to the transporter, who signs and dates all copies to acknowledge receipt of the shipment. One copy is immediately given back to the generator for retention. The other copies must accompany the shipment to its destination. At the destination, the transporter assures that the receiver signs for the waste; he then retains one signed copy and gives the other copies to the destination facility. The facility then returns one signed copy to the generator **within 30 days** and keeps the other for its records.

If you do not receive a signed copy back from the hazardous waste facility within 35 days from the date of shipment, you must contact the transporter or hazardous waste facility to determine the status of that shipment. If the signed manifest still has not been received after another 10 days, you must file an "exception report" with the state agency. This report includes a copy of the manifest and a cover letter describing your efforts to locate the shipment, and their results. **Many states require manifest procedures more complicated than this EPA standard.** For example, **California** and **Massachusetts** require that both the generator and final TSD facility send manifest copies to the state agency within 30 days of sending or receiving a shipment.

If the manifest description differs significantly from the shipment, the TSD facility calls the generator to discuss and hopefully resolve the discrepancy. If the TSD facility operator fails to resolve the discrepancy within 15 days, it must file a **discrepancy report** with the regional EPA office or state agency. This report includes a copy of the manifest and a letter describing efforts to correct the problem.

If you are a large quantity generator, you must certify on these manifests that you are implementing an onsite hazardous waste reduction program, and have selected the safest available treatment and disposal methods for the particular hazardous wastes being shipped.

The federal requirement is less stringent for SQGs. RCRA allows you to ship hazardous wastes offsite without a manifest as long as:

- the waste goes to a recycler who returns reclaimed material to you (the generator)

- the shipments are conducted on a regular schedule under a contractual agreement

- the vehicle is owned and operated by the recycler

- you keep records of the agreement for at least three years after it expires

If a SQG has not received the return manifest from its destination TSD facility within 60 days, the SQG need only send a copy of a manifest to the state agency with a note indicating that the return copy has not been received. Some states are more stringent than this federal practice; for example, **California** allows only a 35-day waiting period regardless of the size of the generator.

Each individual copy of the manifest **must** be kept for a minimum of three years, calculated from the date each shipment was first handed over to a transporter.

RCRA allows hazardous waste generators to accumulate their wastes onsite temporarily, as long as the wastes are handled properly. The time period allowed depends on whether you are a large quantity, small quantity or conditionally exempt small quantity genera-

tor. If you exceed these time limits you are considered to be a storage facility, and must secure a RCRA permit. TSD facility permit requirements are complicated and expensive; it is highly advantageous to most generators to meet the criteria to be "accumulating" instead of "storing."

Under RCRA, large quantity generators generally may accumulate hazardous waste onsite for up to **90 days** without a TSD facility permit. This 90-day period is calculated separately for each container (tank, drum, etc.) and generally begins on the first day on which **any** hazardous waste is placed in that container. However, RCRA does allow for longer accumulation periods under two sets of circumstances.

RCRA regulations allow SQGs to accumulate hazardous wastes without a permit for up to 180 days. If you are a SQG and can show that you ship your wastes at least 200 miles to a TSD facility, you may accumulate hazardous wastes for a long as 270 days. However, you may not exceed total accumulation of 6,000 kilograms of hazardous waste at any one time if you are using this extension. **Many states do not extend normal accumulation time beyond 90 days! New Jersey, California, Washington,** and other states do not provide the 180- or 270-day provisions.

An extension of up to 30 days for the accumulation time limit is available from EPA or the state agency. Extensions are available on an emergency basis when the facility is unable to remove a waste from the site in a timely manner.

RCRA regulations make special provisions for use of "satellite" hazardous waste collection points near active process lines. Accumulation at these satellite stations is **not** subject to the 90-/180-/270-day limits described above. To qualify as a satellite accumulation point, the accumulation point must meet **all** of the following requirements:

- accumulate a maximum of 55 gallons of hazardous waste (or 1 quart of AHW) at a point at or near where the wastes were actually generated, and which is "under the control of the process operator"

- store these wastes in containers other than tanks (e.g., drums) in good condition compatible with the stored wastes, and with hazardous waste labels identifying the wastes and the date accumulation starts.

- transport these wastes offsite within three days of reaching the 55 gallon/1 quart limits. [Alternatively, you may keep (accumulate) the wastes onsite for an additional 90 days from the date the 55 gallon/1 quart limit was first reached. One must add a new date to the label to provide a basis for calculating the new 90-day period, and must follow RCRA and state standards for personnel training and emergency response. If all waste qualifies for satellite collection and is always shipped to offsite within three days of reaching the 55 gallon/1 quart limits, RCRA exempts the facility from these additional requirements.]

This provision is available for any hazardous waste generation that meets these conditions, regardless of how much hazardous waste is generated per month at the facility as a whole. Therefore, one may be able to use satellite accumulation provisions for some

wastestreams even though other wastestreams are subject to the 90-day limit.

As noted earlier, RCRA allows conditionally exempt SQGs to hold wastes for a longer period of time. For AHW, a 90-day maximum accumulation "clock" does not start until you accumulate 1 kilogram of AHW onsite (or 100 kilograms of soil contaminated with AHW). For other hazardous wastes, a 180- or 270-day maximum "clock" does not start until the facility has accumulated 1,000 kilograms. **Some states start the 90-day storage clock at lower accumulation levels.**

A hazardous waste generator must conform to a number of standards when it accumulates hazardous wastes (other than at satellite stations) for any period of time. **These requirements apply fully to large and small quantity generators.** However, as noted above, the facility need not follow these standards at satellite accumulation points if the full containers are removed within three days.

The following are important requirements for generators:

- All hazardous wastes must be kept in **closed** containers that are in good condition and are compatible with the wastes being stored.

- Incompatible wastes are those that react chemically with each other to produce chemical byproducts, heat or other hazards. **Incompatible wastes must not be stored in the same container** nor placed in an unwashed container that previously held an incompatible waste. **Individual containers of incompatible hazardous wastes must be separated** during storage **by a protective barrier** such as a dike, berm or wall.

- All hazardous waste containers must be maintained in a manner that "minimizes the risk" of exposure to the environment. In many cases, this phrase is interpreted to require secondary containment for hazardous waste containers.

- All **containers must be clearly labeled "Hazardous Waste,"** with information included about their contents, accumulation time, generator, and shipping identification. This label must provide:

  - identity of the waste
  - accumulation start date
  - identity of the generator
  - facility EPA ID number
  - hazardous materials information

- All **containers used to store hazardous wastes must be inspected at least weekly** for signs of leaks, corrosion or other deterioration. A generator must maintain (for at least three years) records such as checklists of inspections.

- All generators who accumulate wastes onsite (except those generators who use only satellite storage and then remove their wastes within three days of reaching the volume limit) must develop and implement a **personal training, preparedness, and prevention program** to deal with possible hazardous waste releases and accidents, and must also maintain **contingency plans** and **emergency procedures** to deal with such accidents.

Generators must provide training to familiarize employees with the hazards of the hazardous wastes being managed, and with proper management procedures relevant to their assignments. This program must address both routine handling, and spill and emergency response. These trainings are broader than the Hazard Communication Standard (HCS) training required by the U.S. Occupational Safety and Health Administration (OSHA) and state worker protection agencies, although they can be combined. Employees must successfully complete the program within six months of hiring or transfer to a new position. Only trained employees may work unsupervised. Annual refresher training is required.

Each facility must have its own contingency plan to deal with emergency situations. The plan must be designed to minimize hazards to human health or the environment from fires, explosions or unplanned releases of hazardous waste into air, soil or surface water. The facility must maintain a copy of the hazardous waste contingency plan onsite, and on file with local emergency response agency, fire department, police department, hospital, and the local office of emergency services.

RECORD-KEEPING AND REPORTING REQUIREMENTS

RCRA and state hazardous waste laws impose a variety of record-keeping and reporting requirements on generators. Hazardous waste **generators must prepare and file a report on their hazardous waste activities once every two years.** These reports are due March 1st of all even-numbered years and cover generator activities during the previous year. A copy of the report should be sent to the hazardous waste agency (EPA or the state).

This report covers facility activities during the reporting period. It must include, among other information, the following:

- the facility's name, address, and EPA or state ID Number

- the nature and quantity of each hazardous waste generated, transported or received

- changes in volume or toxicity of wastes achieved through waste reduction

- treatment, storage or disposal methods used for each hazardous waste

- the name, address, and EPA ID Number of each offsite hazardous waste facility to which wastes were shipped during the year

- the name and EPA ID Number of each transporter used during the year

- if wastes received at the facility were generated offsite, the generators' identification numbers

- most recent estimates of the costs for facility closure and postclosure monitoring and maintenance

RCRA exempts SQGs from this requirement. However, some states do require SQGs to report. **Some states (including Washington and Illinois) require** *annual* **hazardous waste activity reports from generators.**

RCRA requires each facility owner or operator (generator, transporter or TSD facility) to keep a **written operating record** at the facility. This operating record must contain the following information:

- the nature and quantity of each hazardous waste generated or handled, along with the method and date of its treatment, storage or disposal

- the location of each hazardous waste generated or held within the facility, and the quantity at each location

- records and results of waste analyses performed

- monitoring, testing or analytical data

- all estimated closure costs

Each generator must keep copies of the following documents for **at least three years**:

- manifests--the three years are calculated from the date the waste was accepted by the initial transporter. The facility should retain both the copy received when it hands over its waste to the transporter, and the additional copy received from the TSD facility, to show that each shipment arrived at its destination.

- results of any waste analyses and tests--the three years are calculated from the date the waste was sent to an onsite or offsite TSD facility

- biennial and other reports--**SQGs are exempt from this requirement.**

EPA or the state hazardous waste agency may extend these periods, routinely or during the course of any enforcement action.

In addition, the facility must keep records related to all personnel who handle hazardous waste, including job titles and job descriptions as well as the names of the employees, their qualifications, and their continuing training. Records must be kept of the required job experience or training given to all facility personnel. Training records on personnel are to be kept for at least three years after the person's termination date, but may accompany personnel transferred within the company.

Source Reduction Requirements

Active pursuit of source reduction is a condition for ongoing compliance with both federal and state laws. Generators must include within their biennial report on hazardous waste activities a report on the progress of their hazardous waste reduction program, including actual changes in the volumes and toxicity of all the hazardous wastes being generated.

When a hazardous waste manifest is signed, you not only certify that the descriptions of the waste are true and correct, you also certify that you have an active waste minimization program in place at your facility. **If you are a large quantity generator**, you certify that you have an onsite program in place to reduce the volume and toxicity of waste generated to the degree you have determined to be economically practicable and that you have selected the practicable method of treatment, storage or disposal currently available to you which minimizes the present and future threat to human health and the environment. **If you are a SQG**, you certify that you have made a

good faith effort to minimize your waste generation and select the best waste management method that is available to you and that you can afford.

Every generator of hazardous waste required to report must include in the biennial report to the appropriate agency on hazardous waste activities information on changes in the volume and toxicity of wastes achieved through waste reduction during the period covered by that report. There are a growing number of laws that require source reduction activities for the hazardous materials they regulate--many of which result in RCRA-regulated hazardous wastes once they are used or discharged.

For example:

- **CWA**--Wastewater treatment systems frequently generate hazardous waste residues

- **CAA**--Air pollution control systems capture noxious compounds that are often then subject to hazardous waste requirements

- **Pollution Prevention Act of 1990** (Section 313 addition to the Superfund Amendments and Reauthorization Act of 1986--SARA) -- Source reduction planning for any of 329 regulated toxic chemicals used onsite above reporting thresholds; most become hazardous wastes after use.

There are also a growing number of states that now implement laws that require hazardous waste generators to take concrete steps to minimize their hazardous waste generation. **Massachusetts** has enacted the most stringent toxics control program in the country; **California** has a mandatory waste reduction analysis and report requirement for large generators, while **Illinois** implements a voluntary waste reduction program.

RECYCLING

RCRA and state laws encourage recycling of hazardous wastes to new uses, as preferable to the treatment and/or disposal of wastes. RCRA regulations consider a waste "recycled" when it is "used, reused (employed as an ingredient in an industrial process or employed as a substitute for a commercial product), or reclaimed (by being processed to recover a usable product)." Resource recovery includes the combustion of high-Btu hazardous wastes in cement kilns, which is considered energy recovery. Frequently, state requirements are more extensive.

Some generators are afraid that recycling activities will make their operations subject to regulation as a "treatment" facility. **Many recycling activities, as well as all "closed unit" treatment operations that occur entirely within an industrial process, are exempt from the definition of treatment (and thus can be carried out without a RCRA TSD facility permit).** In addition, many states implement programs intended to facilitate compliance by many onsite treatment facilities.

In some states, including **California and New Jersey**, a generator who disposes of a hazardous waste that the state considers to be economically and technologically feasible to recycle, may be required by the state agency to submit a written statement justifying why that particular waste was not recycled. After reviewing your statement, the state may order the generator to recycle the hazardous waste by either implementing a feasible onsite method or by shipping the

waste to an offsite recycler identified by the state.

HAZARDOUS WASTE TREATMENT

FACILITIES THAT TREAT, STORE OR DISPOSE OF HAZARDOUS WASTES GENERALLY MUST OBTAIN A PERMIT ("PART B" PERMIT) from EPA or the state agency. This applies to both onsite and offsite activities, and includes onsite hazardous waste treatment. Obtaining a TSD facility permit is a lengthy, complex, and costly process. Permitted TSD facilities are subject to a wide range of operational controls intended to ensure the safety and effectiveness of their operations.

In general, all hazardous waste "treatment" facilities must obtain TSD facility permits. It therefore is important to know what RCRA and the state agency consider treatment, and what exemptions and exceptions may be available. These laws generally provide similar definitions; the RCRA definition states:

> **"Treatment"** means any method, technique, or process, including neutralization, designed to change the physical, chemical, or biological character or composition of any hazardous waste so as to neutralize such waste, or so as to recover energy or material resources from the waste, or so as to render such non-hazardous or less hazardous; safer to transport, store, or dispose of; or amenable for recovery, amenable for storage, or reduced in volume.

RCRA and state laws require that hazardous wastes which are not recycled or recovered be treated before disposal using techniques approved for each wastestream by EPA and/or the state.

A hazardous waste treatment facility is subject to permitting requirements unless it is directly connected to an industrial production process which prevents the release of hazardous waste into the environment. RCRA contains specific exemptions from permitting for various types of treatment facilities; for example:

- elementary neutralization units

- wastewater treatment systems subject to permitting under the National Pollutant Discharge Elimination System (NPDES) or local industrial pretreatment programs

- "totally enclosed treatment facilities" within a process (in this case, no waste is considered "generated")

A few states including **California** and **New Jersey**, have eliminated or modified some of these exemptions.

If the facility treats hazardous waste onsite and it is not exempted from the regulation; **store** onsite at the facility hazardous waste for longer than the accumulation periods described above or store at the facility hazardous wastes generated offsite, or **dispose** of hazardous wastes onsite, then RCRA and state laws require the generator to obtain a permit to do so. RCRA and the EPA regulations refer to these as Hazardous Waste permits or RCRA permits. Permits are obtained through a process typically administered by the state agency with minimal EPA oversight. These permits are also often called "Part B" permits, after the more extensive portion of the permit application

procedure. The permits actually consist of a Part A and a Part B.

Part A permit applications must present at least the following information:

- facility name, mailing address, and location

- design and capacity of the facility

- types and quantities of hazardous wastes managed

- description of hazardous waste management practices

- drawings/photographs of hazardous waste management activities

- EPA ID Number

Part B applications expand this descriptive material to include a detailed description and analysis of:

- hazardous waste analysis, tracking, and manifest system

- hazardous waste packaging, labeling, and transport provisions

- facility standards (covers tanks, drum storage, treatment system design and operation, onsite security, etc.)

- personnel training in waste handling and emergency response

- preparedness and prevention plan covering release response and notification

- contingency plan and emergency response procedures

- onsite inspection program

- record-keeping and reporting systems

- groundwater monitoring (if underground tanks, or land treatment or disposal methods are used)

- closure plan (and provisions for postclosure care, if necessary)

- demonstration of financial responsibility for closure (and postclosure) costs

Once permitted, each TSD facility is subject to extensive onsite self-inspection and record-keeping requirements to document that the programs and procedures defined in its permit are followed. Each hazardous waste TSD facility must prepare and submit, by March 1st of every even-numbered year, a **biennial report** to the state agency or EPA, describing the hazardous waste management activities undertaken at the facility during the previous year. This report resembles the biennial report required of generators. TSD facilities are also responsible for ensuring that generators using their services have undertaken onsite hazardous waste reduction programs. TSD facilities also pay substantial permit fees, and hazardous waste fees to their state hazardous waste programs.

EPA and some states are developing programs to regulate hazardous waste treatment units that can be moved from generator to generator to provide onsite treatment of hazardous wastes, avoiding the dangers of hazardous waste transportation. These units would not be subject to full-scale TSD facility permit requirements, but would operate under a "Permit by Rule" (PBR) system. With a PBR system, the treatment unit operator fulfills limited paperwork requirements, and agrees to operate the unit following all applicable hazardous waste rules. The terminology used to describe these units varies. EPA refers to these units as "Mobile Treatment Units" (MTUs). EPA has studied potential MTU requirements for a number of years, but has no schedule for adopting actual regulations.

While EPA has not yet adopted an official MTU system, one should check with the state agency to find out if it has implemented a similar program or intends to do so in the near future. **California**, for example, has had a PBR program for "Transportable Treatment Units" (TTUs) for several years, covering non-RCRA wastes regulated as hazardous under state law. The California program was amended in early 1992 to provide specific standards for a dozen wastestreams.

SHIPPING WASTES OFFSITE

All generators of hazardous wastes are responsible for their proper shipping, treatment, and disposal, even if they ship the wastes offsite. Generally, all shipments are subject to the **"hazardous waste manifest"** requirements. In addition, specific additional provisions apply to different sorts of offsite activities.

When moving hazardous wastes offsite for storage at another location, the receiving storage facility requires a TSD facility permit. This requirement applies whether you own the offsite facility or it is owned by another company. As a generator, you cannot legally send your hazardous wastes there unless that facility holds such a permit.

If the offsite facility is a "**transfer facility**" -- a location where wastes are kept for no longer than 10 days while they are being transported from one hazardous waste facility (including a generator) to another -- the facility does not require a TSD facility permit. **Some states restrict storage at a transfer facility to less time**; for example, **California** limits time at a transfer facility to 144 hours (six days).

RCRA establishes a series of deadlines by which EPA must define treatment standards sufficient to allow the treated waste residues to be disposed of on the land. Many states have set additional treatment standards for wastes regulated by the state but not EPA.

Federal and state hazardous waste laws call for a nationwide phaseout by 1992 of land disposal of all untreated liquid hazardous wastes in landfills, pits, surface impoundments or similar facilities (referred to collectively as "landfills"). All injection well disposal was banned as of August 1988, except for those few wells determined to be safe for such disposal. Discharges of hazardous substances into water are regulated under the CWA and state water quality acts.

If the facility is a conditionally exempt SQG, RCRA presently allows the facility to mix hazardous wastes with other solid waste and dispose of it at municipal or industrial solid waste facilities that do not hold TSD facility permits as long as total hazardous waste accumulation does not exceed set limits. Not all states recognize this provision; for example, **California** requires **all** hazardous waste to be disposed of at a hazardous waste facility.

INSPECTIONS

EPA and state hazardous waste agencies have authority to inspect hazardous waste generators, transporters, and TSD facilities. RCRA and state regulations require all hazardous waste facility owners and operators to inspect their own facilities for malfunctions, deterioration, employee errors, and discharges which may result in the releases of hazardous wastes to the environment, or pose a threat to human health. The owner or operator must develop and follow a written inspection schedule, and must record all such inspections in an inspection log or summary. Conditionally exempt SQGs are not subject to specific requirements, but still have a general responsibility to manage their hazardous wastes effectively.

The facility's owner or operator must remedy any problems according to a timetable designed to prevent any environmental or human hazards. If a hazard already exists, the facility must take immediate remedial action.

RCRA authorizes EPA to inspect any hazardous waste generator, TSD facility, transporter or underground storage tank (UST) site. EPA can require monitoring, testing, analysis, and reporting to determine whether hazardous waste onsite poses a "substantial threat to human health or the environment." EPA can also sue to address any "imminent hazard." As a practical matter, EPA rarely inspects generators or transporters, focusing instead on selected TSD facilities.

State agencies are authorized to inspect generators and other hazardous waste facilities for state hazardous waste law and RCRA compliance. In some cases, hazardous waste enforcement is conducted out of the regional or district offices of the state agency. County (and a few city) health or hazardous materials safety departments may also take on some inspection responsibilities as part of an overall hazardous materials regulation system.

Under the Comprehensive Environmental Response Compensation, and Liability Act of 1980 (CERCLA), the liability for any problems caused by hazardous wastes is held by **all** responsible parties. Once wastes are generated, the original generator maintains liability throughout these wastes' existence (essentially in perpetuity). This can include:

- all costs of a removal or remedial action incurred by the federal or state government or costs incurred by another person (consistent with the National Hazardous Substances Pollution Contingency Plan, commonly called the National Contingency Plan, developed by EPA and CERCLA)

- damages for injuries, destruction or loss of natural resources, and the cost of processing such damages

CHAPTER 12

MANAGING ENVIRONMENTAL COMPLIANCE

INTRODUCTION AND OVERVIEW

Today's environmental manager must maneuver through 12 major arenas of environmental compliance. Without comprehension of each of these arenas, one can never hope to achieve compliance. Much of the available information about compliance is unfortunately written in either legal or highly technical terminology, making it difficult for straight business managers to comprehend. For this reason, corporations and their facilities require specialists who can read, understand and relate their company's operations to the legal boundary conditions established by the federal programs.

In Chapters 10 and 11, compliance to federal workplace safety and handling hazardous wastes were covered. This final chapter provides an overview of the remaining environmental regulations you must work within. The specific subjects covered are transportation, storage tanks, pesticides, water quality, discharges to sewers, and air pollution.

TRANSPORTING HAZARDOUS MATERIALS AND WASTES

The transportation of hazardous materials and hazardous wastes is regulated under both federal and state laws. The U.S. Department of Transportation (DOT) administers the **Hazardous Materials Transpor-**tation Act (HMTA; enacted in 1981), under which it defines "hazardous materials" (wastes are a subset) and issues inspection, training, and transport requirements. HMTA requirements apply to most forms of transportation including rail, motor vehicles, aircraft, and vessels (pipeline transport is regulated separately). These requirements apply regardless of the destination or origin of the shipment (including transportation between two facilities owned and operated by the same company).

The U.S. Environmental Protection Agency (EPA) regulates transportation of hazardous wastes as part of EPA's administration of the **Resource Conservation and Recovery Act**. In addition, DOT jointly administers with EPA and the Federal Emergency Management Agency (FEMA) a set of incident/spill reporting and response requirements.

States are required to adopt laws and regulations that are consistent with the federal system. Many states have done this by incorporating the relevant federal regulations into their own rules. Some states set additional vehicle and driver safety requirements, and restrict the routes and timing for hazardous materials transportation between states (i.e., inter-state carriers), while state agencies regulate intra-state transport of hazardous materials.

In most states, more than one agency regulates transportation of hazardous materials; and in many cases their authority overlaps. Typically, these agencies include, at a minimum, the state police or highway patrol and the state environmental agency. Other agencies may separately regulate rail transportation within that state.

DOT is changing its transportation regulations to make them more straightforward and consistent with international shipping requirements (following UN recommendations). The new shipping regulations may be followed immediately or phased in accordance with DOT's fixed schedule over several years. Most requirements were to have been met by October 1, 1993, although certain packaging requirements can be deferred until 1996.

By congressional mandate, all other federal and state laws must conform generally to the HMTA. As a result, all hazardous materials transportation nationwide must meet minimal safety standards and common labeling formats. States and localities cannot use overly strict regulations to stifle transportation of these materials.

Under HMTA, **hazardous materials** are defined as those which might create an "unreasonable" risk to health and safety or to property when being transported. Hazardous materials include, but are not limited to, hazardous substances as defined by the federal Comprehensive Environmental Response, Compensation, and Liability Act of 1980 (CERCLA) and hazardous wastes defined by RCRA.

DOT regulations apply both to the **shipper**, who is the person who offers a hazardous waste for transportation, and to the **carrier**, who is the person who transports a hazard-ous material by air, highway, rail or water. As a shipper or carrier of any material appearing on DOT's Hazardous Materials Table (49 CFR, parts 100 to 177) in any quantity or any material listed in DOT's Hazardous Substances Table in a quantity in a single package that equals or exceeds its reportable quantity (RQ) or any RCRA hazardous waste, you are subject to DOT regulations in the following areas:

- packing and repacking

- labeling, marking, and placarding

- handling

- vehicle routing

- manufacturing of packaging and transportation containers

Each of the hazardous materials identified by DOT falls into one or more of the following categories. In 1991, DOT realigned its hazard classes to give them numbers in accordance with international standards:

- Class 1 -- explosives

- Class 2 -- hazardous gases and cryogenic liquids

- Class 3 -- flammable and combustible liquids

- Class 4 -- flammable and combustible solids

- Class 5 -- oxidizers

- Class 6 -- poisons

- Class 7 -- radioactive materials

- Class 8 -- corrosives

- Class 9 -- miscellaneous hazardous materials

DOT has extensively revised its Hazardous Materials Table to incorporate the new information requirements. In cases where more than one type of hazard is present, DOT has established a "precedence of hazard" system so that more severe hazards are addressed adequately.

DOT regulations define documents that must be prepared for each shipment of hazardous materials, as well as labels and placards that must be placed on each shipment. For each material, shipping papers must include:

- DOT-prescribed shipping name

- quantity

- identification number

- the hazard class

Each shipping paper or manifest must contain an emergency response telephone number provided by the shipper, which can be called for additional information in the event of a spill or release.

Shippers must certify that the materials are properly described, packaged, and labeled. For hazardous wastes, the "manifest" (described below) serves as the required shipping document. Manifest forms may be obtained from EPA or state hazardous waste regulatory agencies. Copies of shipping papers must be kept by shippers and carriers for at least six months and may be subject to review by DOT.

The shipper is required to use the most descriptive name available in the Hazardous Materials Table, avoiding use of "not otherwise specified" (n.o.s.) whenever possible for supplementing the n.o.s. description with additional information. In some cases, additional descriptions may be mandatory. For example, if a shipment is made under an exemption, the shipping paper must contain the notation "DOT-E" followed by the exemption number assigned. The words "limited quantity" or phrase "Ltd Qty" must be added for a material offered for transportation under this category; that is, below the maximum amount of a hazardous material for which there is a specific labeling and packaging exemption.

Shippers must ensure that each package is properly marked and labeled and that the transportation vehicle is properly placarded. The carrier cannot accept hazardous materials for transport unless the materials are properly marked and labeled, and the carrier's vehicle properly placarded. There are numerous requirements prescribed for specific hazardous materials, and certain exceptions apply to specific materials or quantities, and may vary by the mode of transport used. The shipper must certify by signature that the materials offered for transportation are "properly classified, packaged, marked, and labelled, and are in proper condition for transportation according to the applicable regulations of the Department of Transportation."

Each package with a rated capacity of 110 gallons or less, and each portable tank and tank car must be marked with the proper shipping name and United Nations (UN) Number or North American (NA) Number as shown in the Hazardous Materials Table. This marking must be on the package surface or on a label, tag or sign. For hazardous substances, the letters "RQ" must also appear on the marking. Cargo tanks and bulk packagings must be marked with the proper identification number on each side and on each end.

Also, each container of hazardous materials must be labeled as specified in the Hazardous Materials Table or the Optional Hazardous Materials Table, unless specifically excepted by the regulations. Generally, the label is a distinctive diamond shape and indicates the hazard class of the materials (e.g., "Nonflammable gas and Poison"). DOT specifies the size, color, and text of each label.

Each vehicle, freight container or cargo tank must be placarded unless excepted. Many hazardous materials have specific placarding requirements. DOT has specified the size and design of these placards, which generally are the same as those of the package labels, and must be prominently displayed on the vehicle. In general, placards are not required for infectious substances, for materials in the D class of Other Regulated Materials (ORMs), for limited quantity shipments, or for properly packaged small quantities of flammable liquids or solids, oxidizers, organic peroxides, corrosives, Poison B, and ORM-A, B, C, and radioactive materials.

ORMs are hazardous materials that either do not meet the definitions of the other hazard classes or have been reclassified as ORM. ORMs are divided into classes: ORM-A materials can cause extreme discomfort to passengers and crew in the event of a leakage; ORM-B materials are capable of causing significant damage to a transport vehicle from leakage; ORM-C materials are specifically named in the regulations and are un-

suitable for shipment unless properly identified and prepared; ORM-D materials are ones such as consumer commodities that present limited hazard during transportation; and ORM-E materials are those not included under other classes, such as hazardous wastes or other hazardous substances.

Certain ORMs may be subject to less stringent shipping requirements (including papers, marking, labeling, and placarding) in the following circumstances:

- ORM-A, B, or C liquid, not over one pint in one package

- ORM-A or B solid, not over five pounds in one package

- ORM-C solid, not over 25 pounds in one package

The DOT regulations also include technical requirements for the construction of containers and transport equipment for each mode of transportation--highway, rail, air, and water. In addition, the regulations outline loading, unloading, and general handling procedures for each mode. For example, liquid hazardous materials must be packed with their closure upward, with markings (orientation arrows, for example) clearly indicating which side is up.

DOT is replacing its cumbersome material-specific packaging standards with general guidance and performance standards based on the characteristics of the hazardous materials being packaged. These standards address such matters as compatibility between the contained materials and the packaging, prohibitions against leaks or emissions, ability to withstand vibrations, temperature variations, etc. The new regulations define **Packing Groups**, provide varying levels of protection, and assign materials to each group as required to provide safe shipping. There are three **Packing Group (PG)** designations that the reader should be familiar with: PG I - High Danger, PG II - Medium Danger, and PG III - Minimal Danger. Materials classified as inhalation poisons must comply with new packaging requirements. Other packaging manufactured after October 1, 1994, must comply with new requirements, and non-complying packaging may not be used after October 1, 1996.

DOT regulates hazardous materials carriers as an extension of its regulation of motor vehicles. The regulations address qualifications for drivers based on physical examination, driving record, and their written and practical driving tests. In **New Jersey**, for example, the Department of Motor Vehicles implements DOT and state requirements covering drivers of vehicles transporting hazardous materials. These drivers must first secure licenses appropriate to the vehicles they drive, and then pass a special examination to obtain approval for hazardous material transport (called a Certified Driver License or CDL). DOT also requires vehicles to meet minimum safety requirements covering lights, brakes, other operating equipment, and maintenance schedules. The 1990 HMTA amendments require certain motor carriers to acquire "safety permits," and require certain transporters of hazardous substances (as defined by EPA) to maintain specified financial responsibility levels.

In some cases DOT regulations may restrict routing of hazardous materials carriers to avoid populated areas, tunnels, and narrow streets when possible. These regulations also require the driver to stop regularly to check tires--every two hours or 100 miles.

Also, each state, often in consultation with city and county governments, establishes route and time restrictions on the transportation of hazardous wastes (and in certain cases, hazardous materials). In general, transportation of hazardous materials must minimize transit time and exposure to congested or dangerous routes.

Effective as of December 31, 1990, whenever there is a shipment of hazardous materials, emergency response information must accompany the shipment. Emergency response information useful in cleaning up a spill must accompany each shipment. The following information must be included for each hazardous material being transported:

- a description and technical name of the hazardous material, as required on the shipping papers

- immediate hazards to health

- risks of fire or explosion

- immediate precautions to be taken in response to an accident

- immediate methods to handle fires

- initial methods to handle spills that do not involve a fire or explosion

- preliminary first aid measures

This information must be available on the transport vehicle, but kept away from the materials themselves; it must be immediately accessible to the transporter and regulatory or fire agency personnel (e.g., in the cab of a truck) for use in responding to a release. Additionally, each "person who offers a hazardous material for transportation" must provide a **24-hour emergency response** telephone number, monitored by personnel able to provide callers with additional information regarding the hazardous materials being transported, and proper emergency response procedures. DOT's regulations provide that this number be monitored while materials are actually being transported, so that those who ship only at certain times need monitor the telephone only during those times.

Carriers must notify appropriate agencies of any transportation incidents (including those which occur during loading, unloading or temporary storage). Generally, immediate notification is required if an incident:

- causes death or serious injury

- involves more than $50,000 in damage

- involves the release of radioactive materials or etiologic agents

- requires public evacuation, or

- causes a major transportation artery or facility to close for more than an hour

In such cases one must make an immediate call to the National Response Center at 1-800-424-8802. Notification should also be made at the same time to the state highway patrol or police.

Notification must include the following information:

- name of person reporting

- name and address of carrier represented by person reporting

- phone number where person reporting can be contacted

- date, time, and location of incident

- extent of injuries (if any)

- proper shipping name, hazard class, UN Number, name, and quantity of hazardous materials involved (if known)

- type of incident, nature of hazardous materials involvement, and whether a continuing danger exists at the scene

The requirements for submitting a written report are more stringent than for making verbal notification. That is, a written report is required even for some incidents that were not sufficiently serious to warrant a call to the National Response Center. DOT requires that its Detailed Hazardous Materials Incident Report (**Form F5800.1**) be submitted within 30 days for each incident where any of the following apply:

- a verbal notification was made to the National Response Center

- any quantity of hazardous materials has been released unintentionally from a package, including from storage tanks

- any quantity of hazardous waste has been released

Generally, you must also send a copy at the same time to the state highway patrol or police. The carrier must retain a copy of the report for at least two years.

These reporting requirements are in addition to similar EPA-administered requirements under RCRA and CERCLA. CERCLA also requires the carrier to report to the National Response Center all spills larger than a specific RQ. [In order to alert drivers and emergency responders to this requirement, the letters "RQ" must appear on all shipping papers if the transporter is carrying in a single container a substance on DOT's Hazardous Substances Table equal to or in excess of its RQ.

If hazardous wastes are released, three additional requirements apply:

- state and local agencies must be verbally informed

- the written report must be made within 10 days

- a copy of the hazardous waste manifest must be attached to the report

- an estimate of the quantity of waste removed from the scene, the name and address of the facility to which it was taken, and the manner of disposition must be included in the report

DOT has incorporated information necessary to comply with these requirements into its *Emergency Response Guidebook*, which can be obtained from the Government Printing Office.

Employees who directly affect hazardous materials transportation safety must receive training in the following:

- familiarization with the regulatory requirements that apply to transportation of hazardous materials

- recognition and identification of hazardous materials

- specific requirements of their job functions involving hazardous materials transportation

Employees who handle or transport packaging containing hazardous materials and those who have the potential for exposure to hazardous materials as a result of a transportation accident must also receive safety training in the following:

- accident prevention

- exposure protection measures

- limited emergency response actions

Drivers must receive training on the safe operation of the motor vehicles they operate. This training must cover the following:

- pre-trip safety inspection

- use of vehicle controls, including emergency equipment

- vehicle operation

- procedures for navigating tunnels, bridges, and rail crossings

- requirements for vehicle attendance, parking, smoking, routing, and incident reporting

- materials loading and unloading procedures

Operators of cargo tanks and vehicles with a portable tank of at least a 1,000 gallon capacity must also receive further specialized training, including retest and inspection requirements for cargo tanks. These driver training requirements may be met through compliance with an appropriate state-issued commercial driver's license that has a tank vehicle or hazardous materials endorsement.

To implement HMTA, DOT is authorized to inspect transport facilities and vehicles as well as records relating to hazardous materials transportation. DOT also can invoke administrative enforcement procedures, including subpoenas and hearings, and issue compliance orders. When necessary, enforcement can be pursued in federal court with assistance from the U.S. Department of Justice. HMTA imposes civil and criminal penalties for violations of its provisions. Amendments to HMTA in 1990 increased civil penalties for violations of its provisions. Amendments to HMTA in 1990 increased civil penalties for violations of HMTA or DOT regulations to $25,000 per violation, with a minimum fine of $250. These amendments also eliminate the defense for an employee who acts without knowledge. Criminal penalties exist for knowingly tampering unlawfully with any document, packaging, label, marking or similar items. Other willful violations can result in criminal penalties of up to $25,000 and/or five years imprisonment. DOT can also seek federal court orders suspending or restricting specific transport activities if the activities pose an "imminent hazard."

As described in Chapter 11, RCRA imposes a comprehensive "cradle to grave" regulatory framework on all activities related to hazardous wastes. This framework includes important controls on the transportation of hazardous wastes. Nationally, EPA has coordinated these requirements with DOT's broader regulation of transport of all hazardous materials. If you ship hazardous waste, you are required by EPA to apply for an EPA Identification Number (ID Number) using EPA **Form 8700-12**. This ID Number must then be used on all documents, allowing federal and state agencies to track each transporter's activities.

All your hazardous waste shipments must be accompanied by a **hazardous waste transportation manifest**. Carriers are to refuse to carry any hazardous wastes that are not documented by manifests. When hazardous wastes are transported across state lines, usually the manifest for the state that will be receiving the wastes must be used. Some states allow the generator-state manifest to be used. A carrier must retain its copy of the manifest for at least three years from the date the waste was first accepted. Carriers must comply with the terms of the manifest. If the shipment cannot be delivered according to the terms of the manifest, the carrier must contact the generator for further instructions and revise the manifest accordingly.

RCRA requires carriers to prepare contingency plans and train personnel in spill response procedures. In response to hazardous waste discharges during transport, transporters must:

- inform appropriate federal, state, and local agencies of the spill

- carry out immediate containment actions (e.g., diking a spill area), although emergency response agencies typically assume site command once they reach the scene of the spill

Carriers retain legal and financial responsibility for cleanup; each responsible party, including the original waste generator, shipper, and/or site or facility operator, also remains potentially liable for the ultimate costs of the cleanup. RCRA's general inspection, record-keeping and enforcement provisions apply to carriers. Thus, a violation of a legal requirement may result in criminal or civil penalties, or both.

Transporters of hazardous materials must turn to regulations for more detailed information on all applicable requirements. These can be found in Volume 49 of the Code of Federal Regulations (49 CFR). The primary source of information specific to chemicals being shipped and their mode of transportation is 49 CFR, parts 171 - 177. Part 172 provides lists of hazardous materials and covers communication regulations while Part 173 covers the shipment and packaging regulations. Part 174 is specific to rail transport, 175 to aircraft, 176 to vessels, and 177 to public highways.

PESTICIDES

Federal law requires comprehensive regulation of the manufacture, handling, and use of pesticides. The U.S. Environmental Protection Agency (EPA), in cooperation with state and local agencies, implements the basic federal regulatory framework governing pesticides known as the **Federal Insecticide, Fungicide and Rodenticide Act (FIFRA)**. This law was initially enacted in 1947 and has been amended several times, most recently in 1988.

FIFRA provides for the registration and classification of pesticides and prescribes controls over their application and use. It allows EPA to delegate primary enforcement responsibility to state agencies. Most states supplement FIFRA's regulatory program with additional state requirements. These requirements fall into four general areas:

- licensing of pesticide applicators and/or dealers

- registration of pesticides with a state agency

- notice and record-keeping procedures

- pesticide storage and use restrictions

Although FIFRA establishes a national basis for the regulation of pesticides, the implementation of FIFRA's regulatory scheme varies from state to state. All states must enforce requirements at least as stringent as those in FIFRA, and many states also have additional requirements. States designate various agencies to regulate pesticides. Depending on the state, the implementing agency could be the Department of Agriculture (as in **Washington** state), the state environmental protection agency (as in **New York** state), or some other agency.

FIFRA requires that each pesticide be registered by use, prior to its distribution or sale. Upon registration, a pesticide is **classified** by EPA according to its use.

Under FIFRA there are three use categories:

- pesticides classified for **general use** are those found not to have unreasonable adverse effects on the environment when used according to directions

- pesticides classified for **restricted use** are those that may have unreasonable adverse effects, including injury to the applicator, if additional regulatory restrictions are not imposed

- the third classification, **mixed use**, is "use" dependent; a pesticide in this category may be classified for general use for some applications and classified for restricted use for other applications

Use classifications may trigger specific federal and state requirements.

FIFRA and state laws regulate the following:

- pest control activities, which are defined to include the use and application of pesticides

- certification of applicators of restricted materials, based on demonstrated competency in the use of pesticides

- worker safety precautions, including requirements to warn field workers before pesticide applications and to prevent exposure of workers not actually involved in pest control activities. EPA and state agencies establish time periods after application during which workers may not return to a treated field.

FIFRA requires registration of each pesticide prior to its distribution or sale. Anyone can register a pesticide, but the pesticide's manufacturer is the most common registrant, since the registration process can take years and be very costly. The 1988 amendments to FIFRA required the **reregistration** of any pesticide containing an active ingredient first registered before November 1, 1984. This reregistration process is still underway. You can ensure that any pesticide product you distribute, sell or use is properly registered by checking the label. The phrase "EPA Registration No." or "EPA Reg. No."

and the registration number must appear on the label of any properly registered pesticide.

The registration process focuses on the proposed use of a particular pesticide, and involves submission of an application which must include:

- name, address, and other applicant information

- complete labeling information and instructions for use

- pesticide name and complete formula

- use classification request

- complete testing data

In addition to this information and other testing information that EPA may require, the application must be accompanied by a registration fee. The amount of the fee depends on the type of pesticide, and ranges from $50,000 to $150,000.

Registration of the following pesticides may be expedited:

- those that would be identical or substantially similar in composition and labeling to a currently registered pesticide ("me too" applications)

- those that would differ in composition and labeling from a currently registered pesticide only in ways that would not significantly increase the risk of adverse environmental effects.

EPA must notify an applicant within 45 days whether the application is complete, and then grant or deny the application within 90 days of receipt. The expedited review process also applies to applications to amend existing registrations that do not require scientific review of data.

An **experimental use permit** may be obtained if additional data are needed before a pesticide can be registered. Any pesticide containing an active ingredient first registered before November 1, 1984, must be reregistered to meet current scientific and regulatory standards. EPA issued four lists in 1989 of those pesticides that require reregistration. **All updated reregistration application information should have been submitted to EPA by October 24, 1990.** If an application for a certain pesticide was not received by that date, EPA initiated procedures to suspend the pesticide's registration.

EPA has 18 to 33 months after its lists are issued to review data submissions to determine whether any gaps in ingredient data exist. Once the data are deemed complete, EPA must decide within one more year whether the pesticide is eligible for reregistration. The registrant then has eight months from EPA's eligibility determination to submit any additional product-specific data required by EPA. EPA then has 90 days to review all data and a total of six months to finalize registration. **Reregistration must be completed by December 1997.**

Two conditions must be met to keep a pesticide registration current:

- the registrant must remain in compliance with FIFRA, applicable EPA regulations, and state law and regulations

- the registrant must pay an annual **maintenance fee** to EPA. The fee depends on the number of pesticide registrations a single registrant maintains, and begins at $650

FIFRA grants EPA the authority to cancel or suspend registrations and to change use classifications. These actions may be taken if new information indicates that the pesticide violates FIFRA provisions, causes unreasonable adverse effects or presents an imminent hazard or emergency. The registrant bears the burden of proof that its registration should not be canceled. Any person who distributes, sells or possesses a pesticide that becomes cancelled or suspended must notify EPA and the state pesticide agency of the quantity of pesticide owned and the location of each storage site. Any person owning pesticide stocks who suffers financial loss as a result of the pesticide's suspension or cancellation may seek an indemnity payment from EPA. This indemnity is not available to a person who, with knowledge of facts indicating that registration requirements were not met, continued to produce the pesticide without notifying EPA.

The 1988 FIFRA Amendments restrict **automatic indemnity** to certain end users such as farmers. With limited exceptions, indemnification for all other persons is not authorized unless congress approves a specific line-item appropriation.

Purchasers of suspended or canceled pesticides who are not end users and who cannot use or resell the pesticides generally can look to the seller for reimbursement. A seller can avoid this reimbursement obligation if, at the time of sale, it notifies the buyer in writing that the pesticide is not subject to reimbursement by the seller in the event of suspension or cancellation.

A **pesticide applicator** is any person who is authorized to use or supervise the use of any restricted use pesticide. Applicators of pesticides classified for restricted use must be certified. Although the general rules for this certification were established by EPA, the states are charged with implementing certification programs. Because of this, the programs vary widely depending on the state. Applicators can be certified as **private** or **commercial** applicators. Private applicators are those who apply restricted use pesticides on:

- their own property

- their employer's property

- another's property, if done in exchange for agricultural services

Commercial applicators are those who apply restricted use pesticides under any circumstances other than those listed for private applicators. They can be certified in one or more categories. EPA's regulations specify 10 categories, but states are free to adopt some, all or additional categories as needed.

EPA's categories for commercial certification are:

- agricultural pest control--further subdivided into plant and animal categories

- forest pest control

- ornamental and turf pest control

- seed treatment

- aquatic pest control

- right-of-way pest control

- industrial, institutional, structural, and health-related pest control

- public health pest control

- demonstration and research pest control

All commercial applicators must demonstrate knowledge in a number of pertinent areas. This knowledge is demonstrated through an examination administered by the state. EPA's regulations define certain basic elements this examination must include; states are free to augment (but not reduce) these basic requirements.

At a minimum, the examination must test the following areas of competency:

- label and labeling comprehension

- knowledge of safety precautions and procedures

- environmental awareness

- pest recognition and relevant biology

- pesticide types, formulations, compatibility, hazards and other relevant knowledge

- equipment types, maintenance, usage and calibration

- application techniques

- applicable state and federal laws and regulations

As with commercial applicators, private applicators must demonstrate knowledge and ability to safely and correctly use restricted use pesticides. This knowledge must be demonstrated by a written or oral examination. At a minimum, an applicant for certification as a private applicator must demonstrate the ability to:

- recognize common pests

- read and understand label information

- apply pesticides according to label instructions

- recognize local environmental concerns that must be considered during application to avoid contamination

- recognize poisoning symptoms and know procedures to follow in the event of mishap

The EPA may require a registrant or applicant to submit information regarding methods for safe storage and disposal of excess quantities of pesticides. EPA may also require that container labels outline directions and procedures for safe transportation, storage and disposal of the pesticide, its container, rinsates, and any other material used to contain or collect excess or spilled quantities of the pesticide. EPA's regulations for FIFRA specify that pesticide transport must comply with requirements set by DOT for hazardous material transport. Also, pesticide dealers are subject to licensing requirements aimed at ensuring safe storage and adequate notice and record-keeping procedures.

FIFRA endows EPA and the courts with broad enforcement powers, including the right to inspect products and records and the authority to prosecute violations and to issue stop-sale orders and injunctions. EPA may order a "**voluntary**" or "**mandatory**" **recall** of a pesticide that has been suspended or cancelled. Voluntary recalls are ordered when EPA determines that such actions will be as safe and effective as a mandatory recall. Under voluntary recalls, EPA asks the registrant to design and submit a recall plan within 60 days. EPA may then approve the voluntary plan and order the registrant to conduct the recall, or may find the plan inadequate to protect health or the environment and order a mandatory recall. Under a mandatory recall, EPA prescribes the terms of the recall which the registrant must implement. EPA can order a mandatory recall plan at the outset or if a proposed voluntary plan is inadequate. The registrant of a pesticide may be required to provide evidence of sufficient financial and other resources to carry out a recall plan in the event of suspension or cancellation of a pesticide. Storage costs incurred as a result of a recall are shared by EPA and the registrant according to a percentage formula.

UNDERGROUND STORAGE TANKS

To protect the nation's groundwater resources, the federal government and most state governments regulate the installation and use of underground storage tanks (USTs). This regulation is implemented on the national level by the U.S. Environmental Protection Agency (EPA), and has three main components:

- registration

- technical standards for construction and operation

- financial responsibility (to ensure funding for cleanup)

State agencies implement state UST laws with varying degrees of coordination with EPA. Most state laws include the three components mentioned above; some states augment the federal program with more detailed requirements.

UST owners register their USTs with a state-designated agency, which in turn reports these registrations to EPA. This requirement is designed to provide EPA and the states with an inventory of USTs, as well as information about their location and size. In New Jersey, for example, registration is through the Bureau of Underground Storage Tanks, a division of DEP (Department of Environmental Protection).

Technical standards are designated to minimize the likelihood that USTs will deteriorate and leak. Existing USTs must upgrade to federal standards on a certain schedule, depending on when they were first installed. New USTs are required to be constructed and equipped to meet strict requirements. State laws often impose stricter standards and tighter deadlines. Also, tank owners must demonstrate financial responsibility to cover the costs of cleanup should a leak or spill occur. EPA has defined (and subsequently revised) schedules of amounts and deadlines to meet this requirement.

State UST laws increasingly follow the federal regulations, often imposing additional provisions covering permitting and use of USTs, more stringent technical standards, and mechanisms to assist owners in meeting financial responsibility obligations. Some states also regulate aboveground storage tanks (ASTs).

Federal law defines a "UST" as any one or a combination of tanks used to store hazardous substances or petroleum products which, including the capacity of connecting pipes, is 10% or more beneath the surface of the ground. The federal UST law defines the following as "hazardous substances":

- gasoline and other petroleum products

- substances for which reportable quantities (RQs) are listed in federal Superfund regulations

- any other substance designated by EPA (none as of July, 1992)

Most states adopt this list into their state UST law, but some states add additional substances. For example, **New York**'s list of regulated substances contains nearly five times as many substances as the federal list.

Tanks exempt from federal UST regulations include:

- USTs, including their piping systems, that are more than 90% above grade

- farm or residential tanks smaller than 1,100 gallons storing motor fuel for onsite use

- USTs storing heating oil for onsite use

- septic tanks

- oil field gathering lines or refinery pipelines

- surface impoundments, lagoons, pits, and ponds

- storm drains, catch basins, and wastewater collection tanks

- separation pumps or flow-through process tanks

- lined and unlined pits

- well cellars

- USTs storing hazardous wastes [these are regulated by the Resource Conservation and Recovery Act (RCRA)]

The EPA required all UST owners to register with their state by May 8, 1986. Since then, new USTs must be registered as they are put into service. Many states have additional registration and/or permitting requirements.

If you removed an UST from service AFTER January 1, 1974, you were required to register these USTs by May 8, 1986. For these out-of-service USTs, you were required to provide the following information:

- the date the UST was taken out of service

- age of the UST on the date it was taken out of service

- size, type, and location of the UST

- type and quantity of any substances remaining in the UST on the date it was taken out of service

USTs removed from service since January 1, 1984, have also been subject to the formal **closure requirements**. EPA requires that you register your UST with your state agency. Some states also require additional steps, which may include the following:

- separate state registration

- permits to install, modify or remove your UST

- payment of registration or permit fees

All the construction and operating standards described earlier apply to USTs defined by federal regulations as "new" or "existing," depending on when they were installed. **In the discussions below, the word "existing" refers to a UST installed before December 22, 1988; the word "new" refers to a UST installed after December 22, 1988.**

Construction standards for "new" USTs include requirements covering corrosion protection, spill and overfill protection, and (for hazardous substance USTs) secondary containment. To comply with "new" UST construction standards, you must meet the requirements in all three areas described below.

The underground portions of all "new" USTs must be designed and constructed to resist corrosion. EPA's regulations define certain options for "new" USTs to meet this corrosion protection requirement:

- the UST is constructed of fiberglass-reinforced plastic

- the UST is steel, and is cathodically protected

- the UST is constructed of a steel-fiberglass - reinforced - plastic composite

- the UST is installed at a "noncorrosive site," as determined by a corrosion expert

"Existing" USTs also have certain options to meet the corrosion protection requirement. One of the following options must be installed:

- a cathodic protection system

- interior lining (such as that described in the American Petroleum Institute's Publication 1631, *Recommended Practice for the Interior Lining of Existing Steel Underground Storage Tanks*)

- interior lining and cathodic protection

Some of these options for "existing" USTs also require monthly monitoring and/or tightness testing.

Unprotected steel USTs often act like a battery. Part of the UST becomes negatively charged and another part positively charged. Moisture in the soil connects these parts and the negatively charged portion of the UST--where the current exits the "battery"--begins to deteriorate. Cathodic protection reverses the electric current. It can come in two forms:

- "Sacrificial anodes" are pieces of metal more electrically active than the UST. Because of this, the electric current exits through them, not the UST, and the UST is protected from corrosion while the anode is sacrificed.

- An "impressed current" system produces an electric current in the

ground from anodes not attached to the UST. This outside current is greater than the current in the UST, protecting the UST from corrosion.

Spill and overfill protection involves both technical and procedural measures. You must both **monitor transfers** into your UST, and **equip your UST with certain hardware**, to prevent spilling and overfilling. Spill prevention equipment is a device to prevent the release of product into the environment when the transfer hose is detached from the fill pipe; for example, a spill catchment basin.

Overfill prevention equipment:

- automatically shuts off flow of product when the UST is no more than 95% full, or

- alerts the operator when the UST is no more than 90% full by restricting the flow or triggering an alarm.

UST systems that store hazardous substances **other than petroleum** must include secondary containment, such as double-walled USTs or external liners, to prevent releases to the environment if the primary container leaks.

Secondary containment systems must:

- contain released product until it is detected and removed

- prevent the release of product to the environment at any time during the operational life of the UST system

- be checked for evidence of release at least every 30 days

Double-walled USTs must:

- contain a release from the inner UST within the outer wall

- detect the failure of the inner wall

External liners must:

- contain 100% of the largest UST within its boundary

- prevent precipitation or groundwater from interfering with its ability to detect or contain a release

- surround the UST completely

Federal law does not require secondary containment for USTs storing petroleum, but many state laws do.

All USTs and piping must be properly installed according to the manufacturer's instructions and codes of practice outlined by nationally recognized associations or independent testing laboratories. For example, one such code is the Petroleum Equipment Institute's Publication RP100, *Recommended Practices for Installation of Underground Liquid Storage Systems.* Many states [including **New Jersey, Pennsylvania, California** and **Ohio**] also require that persons who install or modify UST systems be certified by the state. These certification programs typically involve training in UST installation and repair, and passing an examination.

Whether "new" or "existing," your UST system must provide a method for detecting a release from any portion of a UST that routinely contains product. Your release detection system must have a 95% probability of correctly detecting a leak (and of not falsely indicating a nonexisting leak). Release detection methods that were permanently installed **before** December 22, 1990, are exempt from this 95% standard.

If the UST was installed after December 22, 1988, you are required to implement one of the following two options for leak detection:

1. **Monthly monitoring.** Most UST owners/operators select this option. Any one of the following five methods satisfies the requirements for this option:

   - automatic UST gauging continuously monitors the UST level

   - vapor monitoring detects any vapor from product in the soil above the water table (the "vadose zone")

   - groundwater monitoring detects the presence of any product in or on the groundwater

   - interstitial monitoring (for USTs with secondary containment) detects any product in the space between the primary and secondary containment

   - any other method approved by your state UST agency and EPA can be used

2. **Monthly inventory control *and* UST tightness testing every five years.** This option can only be used for the first 10 years after installation, and these two procedures **must** be used together. Beginning the first day of the eleventh year after instal-

lation, you must switch to monthly monitoring:

- monthly inventory control compares the amount of product that should be in the UST (based on your inventory records) with the amount you find from a monthly test

- UST tightness testing uses equipment capable of detecting a 0.1 gallon per hour leak rate to ensure the integrity of a UST

UST piping can be either **pressurized** or **suction** piping. Standards are generally stricter for pressurized piping, in which a leak would push product out of the piping. A leak in suction piping does not necessarily involve contamination of the surrounding soil. The standards discussed below apply to all USTs. As with other technical standards, some states have different requirements for piping. These differences could include more extensive installation requirements or earlier upgrade schedules.

UST piping must be equipped with leak detection. The differing requirements for pressurized piping and suction piping are discussed below.

If the UST system contains pressurized piping, regardless of whether it is "new" or "existing," you must equip it with leak detection. To meet this requirement, you must chose one method from Group A and one method from Group B, as follows:

Group A methods include:

- an automatic flow restrictor that slows or shuts down the flow of

product through the pipe in the event of a release

- an automatic line leak detector that triggers an audible or visual alarm in the event of a release

Group B methods include:

- annual line tightness testing that can detect a 0.1 gallon per hour leak rate at one and one-half times operating pressure

- monthly monitoring as described earlier--any of the UST methods may be used **EXCEPT** automatic UST gauging

If your UST system has suction piping, regardless of whether the UST is "new" or "existing," you must equip it with leak detection. There are two options you can choose from to meet this requirement:

- line tightness testing conducted every three years

- monthly monitoring--any of the UST methods may be used **EXCEPT** automatic UST gauging

The facility is **exempt** from this requirement and does not need to install leak detection if the suction piping:

- is sloped so that the piping's contents will drain back into the storage UST if the suction is released, **and**

- includes one check valve in each suction line that is located below the suction pump.

As with USTs, piping may also be subject to corrosion. Whether "new" or "existing," the UST piping must be protected against corrosion, using one of two choices to meet this requirement:

- coated and cathodically protected steel piping

- fiberglass piping

"Existing" UST systems also have an additional choice: upgrading the existing steel piping by adding cathodic protection.

If you have a UST that was installed before December 22, 1988 (that is, an "existing" UST), you must **upgrade** it to meet "new" UST standards for leak detection and monitoring, corrosion protection, and spill and overfill protection. Otherwise you must close and remove the UST. Some states have earlier deadlines; check with your state UST agency. Important deadlines for upgrading "existing" UST depends on the date it was installed, as follows:

-- **UST installed before 1965 or age unknown:** December 22, 1989
-- **1965-1969 installation:** December 22, 1990
-- **1970-1974 installation:** December 22, 1991
-- **1975-1979 installation:** December 22, 1992
-- **1980-December 21, 1988 installation:** December 22, 1993

Suction piping must be upgraded as follows:

- Leak detection: same schedule as USTs

- Corrosion protection: by December 22, 1998

Pressurized piping must be upgraded as follows:

- Leak detection: by December 22, 1990

- Corrosion protection: by December 22, 1998

- Secondary containment: by December 22, 1998

Existing USTs must be equipped to prevent spills and overfills from the primary container by **December 22, 1998**.

It is important to understand the response requirements to releases from USTs. A **release** is any spilling, leaking, emitting, discharging, escaping, leaching or disposing from a UST into groundwater, surface water or subsurface soils. Unplanned or unauthorized releases (or leaks) of regulated substances from your UST trigger a series of release recording and reporting requirements.

An unauthorized release occurs when:

- you discover free product at your site

- your equipment is operating unusually

- your leak detection monitoring indicates a release

If your monitoring or inventory control methods indicate that you may have a release, you are given time to attempt to **confirm** the release before reporting. Your time to confirm or disprove a suspected release depends on the type of monitoring you use. If you use **inventory control**, you can wait until your next monthly check. If a second month of data also indicates a release, you must then report the suspected release **within 24 hours**. From that point on, you have seven days to investigate and determine if there has in fact been a release. If you use some form of **release detection**, you can inspect your equipment to determine if it is working properly. If you find the equipment working properly, or repair or replacement of defective equipment does not solve the problem, you must report the suspected release **within 24 hours**. From that point on, you have seven days to investigate and determine if there has in fact been a release.

You must always clean up any release at your UST site. However, some types of releases need not be reported. You do not have to report spills or overfills if:

- they are completely cleaned up **within 24 hours**, and

- they are below the following threshold amounts:

    -- for petroleum products, 25 gallons
    -- for other hazardous substances, the RQ established by EPA under the federal Comprehensive Environmental Response, Compensation and Liability Act of 1980 (CERCLA)

Note that specific states may have different reporting thresholds or require that you report any release.

Once you have confirmed that there has been a reportable release, you **MUST**:

- take immediate action to stop or contain the release

- **within 24 hours** notify EPA and the National Response Center (**1-800-424-8802**)

- notify your state UST agency

Within **20 days after the release**, submit in writing to your state UST agency a report about your cleanup progress and any information you have collected about the release. Within **45 days after the release**, submit in writing to your state UST agency a report containing the findings of your investigation of the release. This investigation should be very thorough, and explore all possible damage or potential damage to the environment. Your state agency or EPA might require additional site studies. Based on your reports, your state UST agency may decide to take further action at your site. You may be required to develop and submit a Corrective Action Plan that shows how you will meet the agency's requirements.

EPA's regulations allow facility owners/operators to repair a leaking UST. The UST repair must be performed according to a standard industry code (for example, National Fire Protection Association Standard 30, *Flammable and Combustible Liquids Code*). Within 30 days after the repair, you must prove the integrity of the repair by one of the following methods:

- have the UST inspected internally **or** tightness tested

- use one of the monthly monitoring methods

- another method approved by EPA or your state UST agency

Within 6 months after repair, USTs with cathodic protection must be tested to check that the cathodic protection is working properly. You must also maintain records for each repair as long as the UST is in service.

Piping made of **fiberglass-reinforced plastic** can be repaired according to the manufacturer's instructions. Within 30 days after repair, you must test the piping using one of the methods explained above for USTs. Damaged **metal** piping **cannot be repaired** and must be replaced.

Federal law requires that owners of USTs storing **petroleum** demonstrate "financial responsibility." Financial responsibility means that the owner or operator must ensure that there will be money available to help pay for the costs of cleanup and possible third-party liability in the event of a leak. The basic requirement is $1 million for each UST containing petroleum, although EPA has lowered and/or delayed this requirement for certain classes of owners. This $1 million requirement does not limit total liability for damages caused by a leak from a UST system--it merely sets a minimum that will allow you to operate a UST.

Owners of the following facilities must have $1 million of per-occurrence coverage:

- facilities with USTs used for petroleum production, refinement or marketing

- facilities that handle an average of **more** than 10,000 gallons per month, based on annual throughput for the previous year

Owners of facilities that handle an average of 10,000 gallons **or less** per month must be able to show $500,000 of per-occurrence coverage. An "occurrence" is a single

accident that results in a release from a UST. The term can have other meanings within standard insurance usage, but in general, "per occurrence" means all the costs associated with cleaning up a single release. Owners or operators must also have enough coverage for annual aggregate amounts to cover all the leaks that might occur in one year. The amount of aggregate coverage depends on the number of petroleum USTs owned or operated.

Financial responsibility can be demonstrated through:

- insurance or risk retention coverage

- a surety bond

- a guarantee

- a letter of credit

- a financial test of self-insurance

- a trust fund

- a state-required mechanism

- a state fund or assurance mechanism

- some combination of the above options

Many states sponsor a state fund or assurance mechanism to assist UST owners in meeting financial responsibility requirements. This may be a fund that you are required to pay into, or it may be deducted from licensing or registration fees. For example, **Ohio** maintains a Petroleum UST Financial Assurance Fund, supported by an annual fee of $150 on each UST. Participation in the fund satisfies federal financial responsibility requirements.

Closures -- USTs may be closed permanently or temporarily. Different requirements apply to each situation. To close any UST permanently, you may have to obtain a UST closure permit from your local UST agency or fire department; in any case, you MUST:

- first, give your state UST agency 30 days notice before emptying the UST of all liquids and accumulated sludges

- then, remove the UST from the ground or fill it with an inert solid material as instructed by your state UST agency. Your closure procedures may have to be monitored by a certified inspector.

- finally, determine if leaks from your UST have contaminated the surrounding environment. You must do this even is you have never had reason to suspect a leak. If your investigation reveals potential environmental damage, you must:

  -- take immediate action to stop and contain the leak or spill
  -- inform regulatory agencies within 24 hours that a leak or spill has occurred (these agencies may vary according to the nature and extent of the release)
  -- remove any explosive vapors and fire hazards
  -- report your progress to regulatory agencies no later than 20 days after confirmation of the leak or spill

Federal UST law defines temporary closure as closure for 12 months or less. While your UST is closed, you must continue to operate the corrosion protection and leak detection systems. If a leak is found, you must respond just as if the UST was operating. You must also cap all lines attached to your UST, except the vent line.

Record-keeping -- Federal UST regulations impose certain record-keeping requirements on UST owners. You need to maintain records in the following areas:

Leak detection: Document your leak detection system's performance and upkeep. These records should include:

- last year's monitoring results, and the most recent tightness test

- copies of performance claims provided by leak detection manufacturers

- records of recent maintenance, repair, and calibration of leak detection equipment

Corrosion protection: Show that the last two inspections of your system were carried out by trained professionals.

Repair or upgrade: Show that any repairs or upgrades to your system were properly done.

Closure: Maintain records for at least three years of the site assessment performed after closure, showing what impact your UST had on the surrounding environment.

An owner who fails to register a UST, or who provides false registration information, is subject to a civil penalty of up to $10,000 per UST. Violations of technical standards or requirements of an approved state program are also subject to civil penalties of up to $10,000 per UST per day of violation.

Failure to comply with EPA compliance orders subjects a violator to a civil penalty of up to $25,000 for each day noncompliance continues. In some states, like **New Jersey**, the fines are higher. EPA can directly enforce federal requirements and bring civil suits in federal court even when a state has obtained EPA approval for its UST program.

WATER QUALITY STANDARDS

The federal Clean Water Act (CWA) provides the basic national framework for regulating discharges of pollutants into the nation's navigable waters. Through the CWA, federal and state agencies establish standards and goals aimed at protecting the water quality in each state. The U.S. EPA has nationwide authority to implement the CWA. States, however, may apply to EPA for authorization to administer various aspects of CWA's program for permitting discharges into waterways. State water quality laws typically parallel the CWA and are designed to allow a state to qualify for delegation of authority to implement federal requirements. Generally, for a state to obtain EPA authorization to administer all or part of the permitting program, it must match the requirements mandated by the CWA or apply more stringent ones.

The CWA is based on a comprehensive permitting scheme known as the National Pollutant Discharge Elimination System (NPDES). The NPDES program requires a discharger to obtain a permit prior to **discharging** any **pollutant** into **navigable waters** from any **point source**. EPA may issue individual permits to dischargers or issue general permits to groups of dischargers. Each of the terms is broadly defined in the CWA to regulate almost all activities that result in the release of contaminants

from a discrete point (e.g., a pipe) and that alter the natural condition of surface water.

NPDES permits establish the level of performance that a discharger must maintain, defined both in terms of the quality and quantity of pollutants. In states that have not received delegation, EPA issues NPDES permits. In states that have received delegation, a state agency issues the permit. Specific states refer to these discharge permits by a state-specific term such as the **New Jersey** Pollutant Discharge Elimination System, or the **New York** State Pollutant Discharge Elimination System.

This regulation defines a **pollutant** as "dredged spoil, solid waste, incinerator residue, sewage, garbage, sewage sludge, munitions, chemical wastes, biological materials, radioactive materials, heat, wrecked or discarded equipment, rock, sand, cellar dirt, and municipal and agricultural waste discharged into water."

As discussed above, EPA may delegate responsibility for administering the NPDES permitting program to individual states that have EPA-approved programs. As of June 1992, 39 states had received EPA approval to conduct state NPDES permit programs. In these 39 states, state water quality agencies issue wastewater discharge permits. In the remaining 11 states, EPA or EPA in cooperation with the state water quality agency issue wastewater discharge permits. Some states have further delegated permitting power to regional or local agencies. For example, in **California** the State Water Resources Control Board (SWRCB) is the EPA-authorized state water quality agency, but NPDES permitting is conducted by nine Regional Water Quality Control Boards located throughout the state.

In order to discharge wastewater from a facility, one must become familiar with the wastewater permitting requirements in the state. Requirements sometimes differ between EPA and the states. Federal regulations mandate minimum nationwide permit application requirements, to which some states have added further requirements. Specifically, federal regulations require the following information be submitted on the permit application at least 180 days (fewer in some states) prior to discharging wastewater from the facility (for new discharges as well as permit renewals):

- the specific pollutants contained in the facility's effluent stream

- the amount or concentration of these pollutants that are discharged into a receiving water

- the location of the outfall (the point source)

Other information that individual states may require in their permit application include:

- the name, address, and telephone number of all current owners and operators of the facility, as well as the name of any parent corporation

- information concerning any administrative action (e.g., consent orders, notices of violation or other corrective enforcement actions) relating to operation of the facility

- a description of the nature of the business at the facility

Wastewater discharge permits must be reviewed and if necessary revised by the facility and the permitting agency.

The CWA describes three broad categories of pollutants:

- **Conventional pollutants** are defined as biochemical oxygen demand, total suspended solids, pH, fecal coliform, and oil and grease. The **best conventional pollutant control technology** (BCT) is required for treatment of conventional pollutants prior to their discharge.

- **Toxic pollutants** include 65 different pollutants identified by EPA. The **best available technology economically achievable** (BAT) is required for treatment of toxic substances prior to their discharge.

- **Nonconventional pollutants** are all other pollutants not classified by EPA and/or your state permitting agency as either toxic or conventional (e.g., thermal discharges). BAT is also required for nonconventional pollutants.

The permitting agency determines the concentration and/or amount of pollutants that may be discharged from your facility on the basis of the established water quality standards for the receiving water and any technology-based effluent limitations or categorical standards or toxic pollutant standards that apply to the discharger.

State **water quality standards** are determined by dividing a state's water bodies into segments, determining the most appropriate use for each segment (e.g., recreational, industrial), setting water quality goals for each segment, and determining what permit conditions to attach to each discharger in order to protect the uses designated for each segment. In addition to water quality standards, permits can incorporate **technology-based effluent limitations** or **categorical standards** established by EPA for specific industries. Categorical standards are established by EPA to provide a national level of pollution control for industrial discharges.

Because it costs less for new facilities to install more advanced technologies than it does to retrofit existing facilities, and because technical alternatives available for new plants may not be suitable for existing sources, Congress requires tighter effluent limits for new sources. EPA has therefore established **national standards of performance** for dozens of specific categories of new sources (electroplaters, textile mills, and manufacturers of organic and inorganic chemicals, among others). These performance standards are more stringent than the technical requirements for conventional, nonconventional, and toxic pollutants, and provide for controls on discharges, both from point sources and into publicly owned treatment works (POTWs). These standards reflect the greatest degree of effluent reduction which EPA determines to be achievable through application of the **"best available demonstrated control technology**, processes, operating methods, or other alternatives, including, where practicable, a standard permitting no discharge of pollutants."

In addition to the national standards of performance, states have the power to provide stricter standards under individual state water quality laws. **New Jersey**, for example, under the New Jersey Water Pollution Control Act, provides its state permitting agency [Department of Environmental Protection (DEP)] with authority to establish New Jersey Pollutant Discharge Elimination System (NJPDES) permit conditions that allow for New Jersey-specific new source performance standards.

In order to identify and control discharges of toxic pollutants, EPA requires states to develop **four lists** describing waters affected by toxic pollutants. By June 4, 1992, states were to have prepared and submitted to EPA an **Individual Control Strategy (ICS)** designed to reduce discharges to toxic pollutants from each point source on the list of point sources preventing improvement of water quality in classified waters (the so-called "C" list). **This ICS is incorporated into the NPDES permit**, the primary control mechanism for reducing point source discharges. Revised permits may therefore be more stringent than previous permits since an ICS must set forth effluent reductions sufficient to attain relevant water quality standards for toxic pollutants. The surface water toxics control program also applies to POTWs. For onsite response actions under Superfund, the ICS will incorporate the decision document prepared for response actions in that area.

There are a number of states that augment EPA's surface water program with their own, often with more restrictive, regulatory efforts. For example, in **Florida**, the Surface Water Improvement and Management Act, enacted in 1987, requires local water districts to prepare a list of water bodies of special significance. The water districts were then to develop surface water improvement and management (SWIM) plans or water bodies. These plans include lists of point and nonpoint dischargers into the priority water bodies, and strategies for restoring and protecting the water bodies.

For owners or operators of an onshore or offshore facility that might discharge harmful quantities of oil into navigable waters, one is required to prepare a Spill Prevention Control and Countermeasure (SPCC) plan. SPCC plans are designed to prevent oil spills, and outline measures that will be taken in the event of a spill. State laws often incorporate SPCC requirements, expanding them to a broad range of facilities. **Pennsylvania's** storage tank law requires that owners of aboveground storage tanks (ASTs) of more than 21,000 gallons prepare spill prevention response plans that are similar in form and content to SPCC Plans. An SPCC plan must be prepared within six months after the facility begins operations, and be fully implemented within one year after it begins operations. The SPCC plan must be amended any time there is a major change in facility design, construction, operation or maintenance which materially affects the facility's potential to discharge oil. In the event of an oil discharge, a copy of your SPCC plan and supporting documentation must be sent to the EPA office in your region and state wastewater permitting agency.

An SPCC plan must be reviewed and certified by a Registered Professional Engineer. It must include a complete discussion of the facility's compliance with construction and operating guidelines, and other effective spill prevention containment procedures including:

- prediction of the nature and extent of an oil discharge that would result from equipment failure

- description of structures and equipment designed to prevent discharged oil from reaching water

- discussion of the facility's compliance with applicable guidelines relating to facility drainage, bulk storage, piping, loading and unloading, oil drillings, facility security,

inspections, record keeping, and personnel training

It is necessary to review and evaluate the SPCC plan at least every three years and amend it (within six months) as necessary, to include more effective, field-proven, and control technologies if they will significantly reduce the likelihood of a spill. In addition, the SPCC plan must be amended whenever there is a change in facility design, construction, operation or maintenance that affects the potential for a discharge.

The CWA contains specific provisions regulating the handling of oil and hazardous substances, and establish specific penalties and rules of liability for the unauthorized release of these materials. These provisions focus on reporting unauthorized (i.e., those without a permit) leaks, spills, and discharges to water. These reporting requirements are similar to those under the federal Comprehensive Environmental Response, Compensation, and Liability Act of 1980 (CERCLA). Any persons in charge of an onshore or offshore facility or vessel or who has knowledge of an unauthorized release of oil or a hazardous substance in a reportable quantity (RQ) must immediately report the release as required by the **National Response Center: 1-800-424-8802**. If notifying the National Response Center is not practicable, one may notify the Coast Guard or the On-Scene Coordinator designated by EPA for the geographic area where the discharge has occurred. To encourage prompt and accurate reporting of spills and leaks, information received through this notification process may not be used against the informant in any criminal case (except prosecution for perjury or for giving false statement). However, failure to provide immediate notification may result in a fine and/or imprisonment.

Groundwater -- The CWA does not regulate the discharge of pollutants into underground waters that do not flow into (are not hydrostatically connected to) surface water. To cover groundwater that is not connected to surface water, the federal Safe Drinking Water Act (SDWA) addresses selected aspects of groundwater protection. Underground injection wells used for waste disposal may pollute groundwater. Underground injection of hazardous waste is generally being phased out under the federal hazardous waste program's "land ban" provisions. To correct contamination of underground sources of drinking water, SDWA regulates underground injection wells. SDWA requires states to develop programs to prevent contamination of drinking water sources by underground injection, and to establish injection well permit programs.

SDWA also establishes a program to protect aquifers (water-bearing strata of permeable rock or soil) from contamination. The Sole Source Aquifer Program provides that any citizen or group can petition that an aquifer provides at least 50% of the water to a community and should therefore be designated as a sole source aquifer. This special protection prevents any land use that receives **any** federal money from proceeding unless a land use plan is submitted to EPA and approved. The Sole Source Aquifer Program does not affect private (non-federally funded) land use. There are currently approximately 50 sole source aquifers in the U.S.

In addition to the Sole Source Aquifer Program, SDWA requires states to develop programs to protect wellhead areas. "Wellhead protection areas" are defined as surface and subsurface areas surrounding water wells or well fields that supply public water systems. States are to implement their wellhead protection programs within two years of submission. Such wellhead protection programs attempt to reduce contamination in the vicinity of the wellhead by controlling the land uses or by requiring liners under certain land uses.

Stormwater Permitting Requirements -- Traditionally, the NPDES program has focused on reducing pollutants in the discharges of industrial process wastewater and municipal sewage. However, the NPDES program has been expanded to regulate a more diffuse source of water pollution: storm water discharges. New federal regulations require industrial facilities that discharge storm water, as well as municipalities with a population of 100,000 or more, to control discharges of storm water and surface runoff and drainage related to storms or snow melt that contain such pollutants as oil, grease and pesticide residue, and flow into storm water drains discharging into the nation's rivers, lakes and streams. These federal regulations apply only to discharges of storm water from point sources; nonpoint sources of storm water are not covered under the new regulatory program.

If a facility is engaged in one of the following "industrial" activities, it must obtain an NPDES storm water permit:

- the manufacturing of certain lumber and wood products, paper, chemicals, petroleum refining, leather tanning and finishing, stone, clay, concrete and glass activities, as well as tobacco, textile, furniture, and printing activities. In short, almost all manufacturing activities may be subject to the new storm water requirements. Facilities involved in these activities are identified in the regulations by Standard Industrial Classification (SIC) code.

- construction activities, including clearing, grading, and excavation (EPA's regulations provided that operations that result in disturbance of less than five acres of total land area and that are not part of a larger common plan of development or sale are exempt from regulation.) This position was recently invalidated in *NRDC v. EPA* and is being reviewed by EPA.

- operation of a hazardous waste treatment, storage or disposal (TSD) facility

- operation of landfills

- operation of a steam electric power generating facility

The federal regulations establish three distinct types of storm water permits: **individual, general,** and **group**. Applying for an individual permit requires submission of **EPA Form 1 and Form 2F**. These forms require dischargers to provide a comprehensive set of information, including:

- a site map

- an estimate of impervious areas

- the identification of significant materials treated or stored on site together with associated materials management and disposal practices

- the location and description of existing structural and nonstructural

controls to reduce pollutants in storm water runoff

- a certification that all storm water outfalls have been evaluated for any unpermitted nonstorm water discharges

- any existing information regarding significant leaks or spills of toxic or hazardous pollutants within the three-year period prior to the permit application

- sampling reports of the facility's storm water taken during "storm events"

The general permit process stipulates regulatory conditions in advance, which dischargers must then meet. This approach allows regulatory agencies to issue a single permit regulating all dischargers of a particular type (which is a more cost-effective approach for permitting than issuing individual permits.) If the facility is regulated under a general permit, it is not required to obtain an individual permit.

The general permit is the procedure most likely to be employed in those states which have general permitting authority. Under the CWA, EPA cannot issue general permits in those state that are authorized to administer the base NPDES program (of the 39 states and territories that have baseline NPDES permitting authority, 11 do not have general permitting authority). Therefore, if the facility is located in one of the 11 states with baseline NPDES permitting authority but without general permitting authority, it has no option except to obtain an individual storm water permit.

In order to apply for a general permit, the facility must submit a **Notice of Intent (NOI)**. The federal regulations provide for minimum NOI requirements. At a minimum, however, the NOI will require the legal name and address of the facility's owner and operator, the facility name and address, the type of facility or discharges, and the receiving stream.

The group permitting option provides facilities involved in the same or similar types of operations with the opportunity to file a single two-part permit application. Part 1 of these applications was due on September 30, 1991. Facilities that missed this deadline are precluded from using the group permitting option and must obtain an individual permit or comply with general permitting requirements (if available).

For purposes of data collection, NPDES permits are required for discharges of storm water runoff from:

- municipal separate storm sewer systems located in incorporated areas with a population of 250,000 or more ("large systems")

- municipal separate storm sewer systems located in incorporated areas with a population of 100,000 or more but less than 250,000 ("medium systems")

Under the federal program, EPA may issue either one system-wide permit that covers all discharges from storm sewers within a large or medium system or distinct permits for categories within a system. Federal municipal storm water permits consist of two parts:

- Part 1 includes source identification information, discharge characteriza-

tions, a description of the legal authority to control discharges, and existing management programs to control pollutants and identify illicit connections to the sewer system

- Part 2 requires additional data--a proposed management program, an assessment of storm water controls, and fiscal analysis

Publically Owned Treatment Works

The federal Clean Water Act (CWA) regulates all direct discharges into navigable waters. All such discharges, including treated sewage discharged by sewer systems and sanitation agencies (publicly owned treatment works--POTWs), require permits issued under the **National Pollutant Discharge Elimination System (NPDES)**. These NPDES permits restrict the quantity and quality of a POTW's discharge. To meet these permit requirements, individual POTWs place a range of detailed restrictions on their own industrial users, subject to selective federal, state, and regional oversight of such "pretreatment programs."

EPA has overall responsibility for the CWA, including development of selected national pretreatment standards for pollutant discharges into POTWs. States that are authorized by EPA to administer the federal pretreatment requirements, will administer this program through its state water quality agency. POTWs obtain permits for their discharges from EPA or the state water quality agency by following essentially the same process required of direct industrial dischargers. POTWs then regulate industry discharges into their sewer system to be sure of meeting their own discharge limits. This regulation by POTWs of "indirect dischargers" to the treatment works results in environmental regulation that is uniquely tailored to local conditions. The federal government, however, has not completely abdicated its regulatory role. EPA sets general pretreatment standards and national categorical pretreatment standards for different industry categories, defining necessary effluent reductions. EPA or authorized states may also place more stringent requirements on POTWs if necessary to meet water quality standards for the receiving waters. To meet these standards, POTWs may in turn regulate indirect dischargers.

Industrial users of POTWs are not required to obtain NPDES permits and are ordinarily not required to obtain a state-issued discharge permit. Instead, their individual POTW imposes restrictions--"pretreatment standards"--on all its industrial users. Typically, a POTW discharge permit is required from the POTW, although POTWs differ in their terminologies.

Regulation of industrial discharges into POTWs serves four objectives:

- prevents introduction of pollutants that would prevent the POTW from complying with its own NPDES permit and discharge limitations

- prevents introduction of pollutants into POTWs which would interfere with equipment or operations, or endanger POTW personnel

- prevents introduction of pollutants that would "pass through" (i.e., would not be treated adequately before discharge to the receiving water body) or would be incompatible with the POTW

- improves opportunities to recycle and reclaim municipal and industrial wastes and sludges

In order to meet these objectives, each POTW imposes discharge limitations on some or all of its industrial users; POTWs also enforce relevant federal or state pretreatment standards.

EPA's general pretreatment regulations provide for the following controls over POTW dischargers:

- POTW evaluation of "significant industrial users" to determine if additional controls are required to prevent "slug" discharges

- inspection of the discharger by the POTW, at least annually

- submission of sampling and enforcement response plans by dischargers without categorical standards.

- required reporting of any discharges to a POTW of any RCRA hazardous wastes not subject to self-monitoring requirements

EPA has established a variety of national pretreatment standards that all facilities in a particular industrial category must achieve before discharging their effluent to a POTW. POTW's pretreatment programs enforce these national standards, as well as their own additional local requirements. An industrial user in any of the designated categories, must conform to national **categorical pretreatment standards** for existing and new sources (or seek a variance from these standards). These pretreatment standard specify quantities and concentrations of pollutants that may be discharged into POTWs and require dischargers to use "**best available technology economically achievable**" **(BAT)** prior to discharging.

The CWA requires POTWs with a design flow **over five million gallons per day** (mgd) which receive from industrial users pollutants that pass through or interfere with the operation of the POTW to establish industrial pretreatment programs to control industrial discharges into their sanitary sewer systems. POTWs with a design flow of five mgd or less may also be required to establish pretreatment programs if the nature or volume of industrial discharges to the POTW, upsets, violations of POTW effluent limitations, contamination of municipal sludge or other circumstances pose a danger of interference or pass through. EPA or the state water quality agency enforces this federal program statewide. Other POTWs may also choose to establish local pretreatment programs.

Local standards are often more stringent than the national standards, or cover industry groups not covered by the EPA standards. These local requirements are often imposed in response to unique concentrations of point or nonpoint discharges into the receiving waters, or to provide additional protection to receiving waters. In many industrial areas, pretreatment programs provide important elements of the area's overall toxics management effort.

POTWs with approved pretreatment programs are required to issue permits or equivalent individual control mechanisms, and to inspect and sample each significant industrial user at least once per year. Other POTWs issue permits on their own, separate from the federal requirements.

Pretreatment controls may include individual industrial use permits (as discussed above), permits-by-rule, or the prohibition of certain discharges to the sewers. Permits issued by the POTW or permit-by-rule can establish flow restrictions, pollutant concentrations or other pretreatment thresholds and set forth specific monitoring and reporting requirements.

POTWs with pretreatment programs are required to implement an enforcement response plan with procedures to investigate and respond to industrial user noncompliance. Many POTWs conduct regular inspections of their industrial sewer users.

RCRA excludes solid or dissolved material in domestic sewage (i.e., POTWs serving residences) from its definition of "hazardous wastes." As a result, if you operate an industrial facility that is permitted by your POTW to discharge "hazardous" liquids into sewers, you are not subject to RCRA manifest requirements even though these discharges would otherwise be considered hazardous waste (this is known as the "**domestic sewage exclusion**"). However, you **must still comply** with any other applicable hazardous waste management requirements (e.g., obtain an EPA Identification Number and comply with both RCRA and state hazardous waste law standards for treatment and storage).

As noted, such discharges must also meet the POTW's specific pretreatment requirements. POTWs which receive pretreated hazardous liquids along with domestic sewage are not considered to have received hazardous waste, and therefore are not required to meet any of the treatment, storage, and disposal requirements mandated by RCRA.

An industrial user, is required to submit written notice to POTWs, EPA, and state hazardous waste authorities of any discharge that would be considered hazardous if not made under the domestic sewage exclusion. These notifications must contain the name and EPA hazardous waste number of each hazardous waste, and the type of discharge (continuous, batch or other). If the discharge is more than 100 kilograms per month, the notification must also include: identification of the hazardous constituents; estimates of the mass and concentrations of these constituents discharged monthly; and estimates of the mass of discharges in the coming 12 months. Notifications are due within 180 days after the discharge. Notifications need only be filed once with EPA and the state for each hazardous waste being discharged, but must provide the POTW with a follow-up notification whenever there is a change in the discharge. Federal regulations provide for specific exemptions from this reporting requirement.

AIR QUALITY STANDARDS

The federal government and individual states regulate literally thousands of commercial, industrial, and other activities to clean up the nation's air. Strong amendments were made to early versions of the **Clean Air Act (CAA)** in 1970, again in 1977, and most recently in 1990. The law requires the EPA to set health-based national standards for several different pollutants.

Within the framework provided, air quality regulatory compliance varies greatly from one state to another, and even within a state. While all areas are responsible for meeting the same basic national air quality standards, facilities in certain locations may have to comply with more stringent requirements if their air quality is significantly worse (or even significantly better) than the national standards. State and local agencies are given a great deal of discretion in implementing the nation's air quality laws, as well as additional state and local requirements.

In nearly every state, three layers of governmental agencies are involved in regulating air pollution. EPA sets national standards and oversees selected state and local agency actions. The **state air quality agency** continually revises each **state implementation plan (SIP)**, enforces federal standards, controls auto emissions, and sets guidelines for local air pollution control agencies (if any). Many states set and enforce standards for some pollutants which are more stringent than the national standards. Actual permitting is typically performed by **local** agencies--typically state agency field offices, or in some states, regional or county air quality control authorities. These agencies prepare local plans, establish local rules, issue permits, and enforce the relevant requirements through inspections and other procedures.

These programs distinguish in important ways between **existing sources** of air pollution, and **new sources**. New source requirements are often more stringent, on the theory that pollution control technology can be built into new plants from the ground up, and that new sources add to existing air pollution levels. This focus on stricter technology controls for new sources applies to both stationary and mobile sources of air pollution.

Conventional and toxic pollutants are addressed differently. While conventional air pollutants are present in greater quantities, they are also less dangerous, pound for pound, than are air toxics. Both kinds of air pollutants can harm public health.

The 1990 CAA amendments require more emissions inventories from smaller sources than ever before, and also target these polluters for accelerated emissions reductions depending on their area's non-attainment status for the different pollutants--extreme, severe, serious or moderate. Many of these sources will fall under a new federal air quality permit program that began in the mid-1990s. The 1990 CAA amendments also tightened transportation control measures, and imposed the stiffest civil and criminal penalties yet seen for environmental crimes.

Existing sources of air pollutants are regulated according to their emission, sources, and quantities. Provisions vary based on location. Some local agencies often impose additional requirements.

Six "criteria" or "conventional" pollutants are regulated under CAA:

- carbon monoxide (CO)

- lead

- nitrogen oxides (NO_x)

- ozone (smog)

- particulates (10 microns or less in size, or "PM-10")

- sulfur oxides (SO_x)

Several of the conventional air pollutants are familiar as the building blocks of urban smog. Others pose additional problems for health and visibility.

Some facilities emit smog precursors even though they don't have smokestacks. For example, large bakeries in many areas must install control devices on their ovens to reduce the amount of ethanol--formed by yeast when bread dough rises--which they emit into the air. Ethanol is a **Volatile Organic Compound (VOC)**. Many VOCs are also **Reactive Organic Gases (ROGs)** which can contribute to smog formation. Thus, conventional air pollutants are not just as the six listed "criteria" pollutants, but also as a broader array of pollutants that interact with (or form) these six gases to produce smog, haze, and acid rain, for example.

EPA has set **National Ambient Air Quality Standards (NAAQS)** for the six criteria pollutants, defining air quality targets. When a region or air basin has not achieved these standards in its background (or "ambient") air quality for one or more of these six pollutants, that region is said to be a "**non-attainment area**." Air quality agencies in these areas must impose additional and more stringent or innovative measures to improve air quality toward NAAQS compliance, especially under the requirements of the 1990 CAA amendments.

PSD (Prevention of Significant Deterioration) standards focus on maintaining air quality in especially pristine areas. PSD standards apply to SO_x, particulates, and NO_x, but not to ozone or smog itself. These standards were mandated by Congress in the 1977 CAA amendments to reflect the fact that some areas with pristine air quality should be afforded an extra measure of protection. PSD standards fall into three categories: Class I, Class II, and Class III.

PSD Class I areas automatically include:

- International Parks

- National Wilderness Areas and National Memorial Parks over 5,000 acres in size

- National Parks over 6,000 acres in size

These standards allow very little additional emission of certain air pollutants from certain kinds of large new (or modified) stationary sources (such as power plants) proposed in or near such areas. As a result, it is very difficult to obtain approval for such large new sources of air contaminants in PSD Class I areas, or nearby.

PSD Class II and Class III standards apply in areas designated primarily by the states; usually scenic rivers and some wilderness or agricultural areas are Class II, whereas urban areas are usually Class III. The air in Class II and Class III areas is generally cleaner than the air in areas that are barely in attainment with NAAQS.

Most states have had their own air emission permit requirements for many years. The 1990 CAA amendments introduced for the first time a **new nationwide air quality permit program**, to take effect by the middle of the decade. In most areas, federal air quality permits will be issued by today's state or local permitting agencies; however, these permits will contain many new features mandated by the 1990 CAA amendments.

Many **stationary sources of air pollutants** need a permit issued by the appropriate air quality agency; details on which facilities need a permit vary considerably from one

area to another. Most air quality agencies also require that facilities obtain a permit to **install and use any air pollution control device**.

The thresholds for emissions that require permits are being lowered in order to bring many small sources under air quality rules and regulations that previously applied only to large sources of air pollutants.

The following are categories of equipment and processes for which permits are often required by air quality agencies in non-attainment areas:

- Combustion

  -- Boilers, heaters, furnaces, turbines
  -- Project-associated combustion (cogeneration, resource recovery)
  -- Internal combustion engines

- Open-air Processes and Operations

- Metal Melting

- Incineration

- Material Processing (and Equipment)

  -- Coating
  -- Cleaning
  -- Food Processing
  -- Drying
  -- Curing
  -- Printing

- Production of Chemicals, Petrochemicals, and Petroleum Products

- Handling of Grains, Feed, and Food Products

- Handling of Minerals

- Storage and Handling of Natural Organic Materials

The term "**control technology**" simply refers to any piece of equipment added to another piece of equipment or machine solely for the purpose of controlling (i.e., reducing) emissions of air contaminants. Some control technologies, such as the catalytic converters used on automobiles, are built into equipment right from the design stage; others are added after a machine or industrial process has been operating. Other typical control technologies address the type of equipment or methods that must be used to control emissions.

Air districts often require that boilers and steam generators be designed or retrofitted to emit low levels of NO_x. The most common control technologies for boilers are low-NO_x burners, flue-gas recirculation, selective catalytic reduction, and selective noncatalytic reduction.

Air quality agencies in non-attainment areas typically require a two- or three-step process for permits for equipment and processes that emit air contaminants. The first step usually requires sources to obtain a **permit to construct or install** a piece of equipment, or to build an entire facility. Sources at this stage must obtain a "Permit to Install," "Authority to Construct," or "Permit to Construct." These different terms are used interchangeably in the various state air pollution control programs. In the second step, following construction, a source must obtain a "**Permit to Operate**" or "**Authority to Operate**." Occasionally, the permit to construct may serve as a temporary operating permit, whose duration is limited to one or two years. The third step involves **periodic permit renewals**.

Most air districts assess annual permit fees to cover the costs of their review of facilities and permit applications, and their enforcement activities. Fees are assessed on a per-equipment basis. They vary with the size and complexity of the equipment. Typically, air districts send permit holders invoices for their air quality fees, either on the anniversary of the date their permit took effect or on a date triggered by that air district's rules and regulations.

At a minimum, permits under the federal permit program must require:

- enforceable emissions limitations and standards

- a compliance schedule for each individual facility

- progress reports (including air monitoring results) to be submitted every six months

- reports of any deviations from permit requirements (to be submitted immediately)

- facility record keeping

State and local air quality agencies develop administrative and source- or industry-specific rules. The complexity and comprehensiveness of these rules vary a great deal. Generally, agencies located in areas with poor air quality tend to be more comprehensive in their rule-making, as they struggle to meet NAAQS. Air quality agencies write two kinds of rules in both attainment and non-attainment areas: administrative rules and source-specific rules. **Administrative rules** govern the interaction and paperwork between the air quality agency and the operator of a controlled source of air pollutants. Administrative rules describe who needs a permit, how the permit should be applied for and obtained, how long it will be valid, and where it should be posted. These rules also outline the agencies' responsibilities toward source operators in their jurisdictions, including how many days they have to respond to petitions by sources, when they can inspect a source and issue penalties, and grievance procedures. **Source-specific rules** cover certain industrial or commercial processes and equipment. These rules are usually written narrowly enough to target emissions of one kind of pollutant from one set of similar processes or equipment. They often include a mix of equipment standards, materials standards, and procedures (or "process standards") describing how equipment or a procedure can be used to minimize emissions. A rule typically defines how many pounds or tons of a certain pollutant can be emitted, per day or per manufacturing process. Alternatively, the rule may specify that a business must use a specified process or products which do not contain more than specified amounts of pollution-causing chemicals. Rules may become more stringent over time.

All air districts define **limits** for various emissions as required to meet SIP goals and the targets set out in regional or local air quality plans. A limit is the maximum emission allowed from a particular kind of source. It is usually developed balancing health and other air quality concerns with technological and economic feasibility. In heavily polluted non-attainment areas like southern **California**, limits tend to be more stringent and exist for a greater number of pollutants. These limits are typically set out in numerous rules, each covering a particu-

lar industry category, pollutant or type of activity.

Air districts often go beyond merely mandating daily or annual emissions totals to dictate **performance standards** for equipment used in the processes that generate air emissions. For example, a number of agencies now require businesses that offer spray coating services to use an HVLP spray gun. By using less paint, the operators may save money over the life of the equipment. As new regulations are passed, some businesses occasionally find that previously unregulated equipment now falls under a new inspection process. A permit may even specify in detail what equipment should be used.

Air quality laws provide a number of "technology forcing" mandates to agencies to devise technical performance standards for emission controls. The laws have coined a number of terms for these various sets of standards. Air districts may be required to apply **Reasonably Available Control Technology (RACT)** in some cases, and **Best Available Retrofit Control Technology (BARCT)** in others. Occasionally, state air agencies or air districts may have to apply even more stringent **Best Available Control Technology (BACT)** or mandate use of the **Lowest Achievable Emissions Rate (LAER)**. The principal differences among these terms have to do with the level of air pollution control technology required, the air quality designation of the jurisdiction for which the control measure is chosen (moderate, serious, sever non-attainment), and the "impact" of controls--a combination of environmental, economic, and energy considerations. Thus, some control measures are more purely "technology forcing" (requiring firms to use state-of-the-art, expensive control equipment), while others balance gains in air quality against equipment costs. Energy requirements are also factored in because most control equipment adds further energy demands to industrial or heat-producing processes. Extra energy, of course, often means more emissions from fossil fuel combustion.

Sources may obtain variances from certain air district rules and regulations. Variances are temporary exceptions from administrative laws and rules intended to give firms the necessary "breathing space" to continue operating while they take the steps needed to meet specified air pollution control requirements. **Variances are granted only for permits to use or operate a source, and for rule compliance actions. Variances cannot be granted for permits to build, erect, alter or replace a source.** Persons seeking a variance usually must petition their air quality agency. Often, these petitions are reviewed by a Hearing Board made up of appointed citizens. This Hearing Board has authority to deny or revoke permits, approve or reject new or proposed rules, and grant variances from district rules. Each Hearing Board has its own quasi-judicial administrative procedures (i.e., courtroom-like) for conducting reviews and hearings.

Federal rules define **"new"** or **"modified"** sources as those facilities that **either** add further air pollutant emissions to a site, **or** undergo construction costs equal to 50% or more of the costs of building comparable facilities from the ground up. Thus, it takes relatively little modification for a facility covered by these rules to come under federal new source requirements. Generally, the term "new sources" is used to refer to both new and modified sources. New sources of air pollutants typically face more stringent controls than do existing sources, on the presumption that it is easier to incorporate

new control technologies into the design stage of a new facility than to add them later to an existing facility. Details vary greatly depending on the local area, the type of facility, and the size of the facility.

New sources may have to go through a permitting process called **New Source Review (NSR)**. In this process a state or local air quality agency determines whether a proposed or modified facility can begin operations, and under what conditions. Since state and local requirements are often more stringent than federal standards, you will need to check with your local air quality agency to determine your requirements and status. The NSR process generally applies to major new sources. Most smaller new sources can proceed through the normal permitting procedures without triggering the NSR process. EPA generally defines a major source as one that emits 100 tons or more per year of any air contaminant or mix of air contaminants. Where local or state rules do not specify their own definition of "major sources," sources defined as "Major Polluting Facilities," or Major Modifications of Major Polluting Facilities must meet federal requirements. However, many state and local agencies apply lower thresholds. Different emission thresholds trigger NSR in different non-attainment areas, based on how far the areas exceed NAAQS. The most stringent federal NSR requirements apply to new sources that emit 10 tons or more per year of any air contaminant. This threshold applies only in southern **California**'s South Coast Air Basin. NSR thresholds for other non-attainment areas around the country range from just over 10 tons to 100 tons per year of any pollutant. Some air districts covering non-attainment areas enforce local rules more stringent than federal regulations. For example, most air districts covering non-attainment areas in EPA's Region 9 have NSR thresholds for some pollutants or processes that are less than the 10 ton federal threshold.

The CAA requires EPA to develop NSPSs for certain categories of new sources located anywhere in the country. These standards require the best technologically feasible system of emissions reduction for each category, provided those controls are economically viable. A typical NSPS sets out the date upon which it becomes effective, the kind of process or equipment covered, and requirements that must be observed by the new source operator. For example, the NSPS for new bulk gasoline storage tanks describes storage tank and tank roof designs that all your new gasoline tanks larger than 40,000 gallons (and containing certain liquids) must meet. The basic idea of an offset system is to allow one source to "compensate" for its own emissions--especially its new or increased emissions--by reducing the pollution levels of other sources in its same airshed. This compensation could involve buying some businesses that are heavy polluters, and closing them while retaining their right (under earlier permits) to emit pollutants. It could also mean installing control devices on other sources that result in emissions reductions at least as great as your increase. By letting sources "shop" for offsets, these programs are intended to reduce the costs of compliance. Some offset programs require new or modified sources in polluted areas to reduce emissions by producing offsets larger than the emissions they will create. For example, under the 1990 CAA amendments new sources in extremely polluted areas such as Houston or Baton Rouge will be required to eliminate 1.5 tons of pollutants for each ton they expect to emit. This system ensures that introduction of a new source into the area will actually result in a **net reduction** in emissions, instead of an increase or even no net change.

Many state environmental quality acts require new projects to be evaluated through an **Environmental Impact Report (EIR)** or similar environmental assessment process if the project may present significant environmental effects. EIRs can be required even if only one environmental medium (e.g., air) would be affected by a proposed project. Completion of an EIR is not a substitute for obtaining an air quality permit; the EIR process operates **in addition** to required state agency or local air district permit and rule requirements, including NSR requirements.

"Market incentive" tools in the battle against dirty air include emissions credits, marketable permits or emissions trading systems. These systems allow companies to meet emission reductions as they see fit-- either by installing air pollution control equipment or by purchasing "credits or permits to pollute" from other companies. In an emissions trading program, companies that exceed their emissions targets or emit less than their allowance may sell their surplus "emission credits" in an open market. These programs make broader use of the ideas and mechanisms developed for so-called offset programs. Title IV of the 1990 CAA amendments sets out a massive new Acid Rain Control Program using a marketable allowance system. The aim is to reduce and control SO_2 and NO_x emissions from fossil fuel-fired electric utility plants. SO_2 and NO_x emissions from power plants react with atmospheric gases to form acid rain, which can be deposited hundreds of miles away from the original source of the emissions. This program covers utilities except those with $\leq$ 25 MW capacities, qualifying cogeneration units, and small power producers. Under Title IV, EPA is required to allocate tradeable SO_2 emissions allowances among major sources. Each allowance is equal to one ton of SO_2. EPA will issue to each particular source a number of allowances equal to the annual tonnage emission limitation set out in its permit. If a source reduces its emissions below its allocation, it may sell its extra allowances to other sources. New sources will need to purchase allowances for their SO_2 emissions from existing sources. Existing sources may also choose to do so if these purchases cost less than reducing their own emissions directly. An example of a simplified calculation sheet for recording the VOC (called VOM in **Illinois**) emitted in a typical metal coating operation is given on the next page.

Most of the air quality regulations discussed thus far concern stationary sources of pollutants. These sources have traditionally borne the brunt of air quality control measures because, unlike individual mobile sources, they are relatively large and identifiable, and they don't move from one jurisdiction to another on a daily basis. Mobile sources cannot be ignored, however, because they represent a very large cause of air pollution. In the urban areas of **California** that are still out of NAAQS attainment for ozone and smog, for example, on-road motor vehicles are responsible for 37% of ROG emissions from all sources, 51% of all NO_x emissions, and 69% of all CO emissions. There are two major ways to control emissions from mobile sources. The first is to change the technologies we use for transportation; the second is to control the indirect sources associated with automobiles. Changes in transportation technology involve improvements in fuel efficiency and emission controls on motor vehicles, or the use of "cleaner" transportation energy such as electricity. Several states, including **Colo-rado, Massachusetts,** and **California**, have taken steps to mandate cleaner fuels and technologies. The **Federal Motor Vehicle Control Program (FMVCP)** consists of a series of regulations applicable to mobile sources such as automobiles, trucks, aircraft, and other mobile equipment that emit air contaminants. Mobile source emissions are covered by emissions standards (defining how much of any given pollutant may be emitted), equipment standards (e.g., catalytic converters), and materials standards (composition of motor fuels). Some states (especially those with particularly poor air quality in metropolitan areas) implement emission testing programs. Metropolitan areas with more than 200,000 people were required by EPA to incorporate smog testing into their SIPs.

The 1990 CAA amendments mandated a number of new mobile source programs. For example, **California**'s existing clean fuels requirements have been redefined as a pilot program for the rest of the country. In 1988 the city of Los Angeles decided it would purchase 15,000 "clean-fuel" vehicles for use by various city agencies. By 1996, 150,000 clean-fuel vehicles must be made available for sale in California; by 1999, this figure will increase to 300,000 vehicles per year. **Colorado** offers a $200 incentive to alternative fueled vehicle owners who certify their cars with the state.

In 1992, EPA proposed rules regarding on-board emissions diagnostic systems that may replace some existing inspection and maintenance tests. In 1994, all new transit buses are required to use clean fuels like methanol, natural gas or electricity. By 1995, cleaner, reformulated gasoline must be produced for sale in the nine areas with most severe ozone pollution across the country. Also by 1995, all new vehicles

VOC CALCULATION FOR METAL COATING OPERATION

A metal coating operation applied 0.53 gallons of coating in one day. The make-up of this coating is as follows:

amount of VOC in the coating
(less water and exempt compounds)   3.5   lbs/gal
amount of VOC in the reducer (hardener)
(less water and exempt compounds)   6.625   lbs/gal

mix ratio of coating to reducer   2:1

To determine the VOC value of the mixture:

Multiply the amounts of coating and reducer by their proportions in the mix:

coating	3.5	x	2	=	7 lbs/gal
reducer	6.625	x	1	=	6.625 lbs/gal

sum   13.625 lbs/gal

Add the number of parts in the mix to determine their total:

$$2 + 1 \; = \; 3$$

Divide the sum of the amounts of coating and reducer by the number of parts in the mixture:

$$13.625 \div 3 \; = \; 4.54 \; \text{lbs/gal}$$

The VOC value of the mixture is 4.54 pounds per gallon.

To determine the VOC applied that day:

Multiply the VOC value of the mixture by the amount of mixture used:

$$4.54 \; \text{lbs/gal} \; x \; 0.53 \; \text{gal} \; = \; 2.4 \; \text{lbs}$$

The total daily VOC emission is 2.4 pounds.

will be required to have on-board vapor recovery canisters.

Concern over emissions from mobile sources has also spawned some controls on stationary sources associated with automobiles. For example, gas stations in non-attainment areas must equip their pump nozzles with vapor lock and recovery systems that reduce VOC emissions from filling operations.

CAA and state air laws all give air quality agencies authority to enforce their requirements. Most state laws empower air districts to send inspectors into any site, facility, building or equipment at any time to determine compliance with air pollution control laws. In **New Jersey** this also includes the right to test or sample any materials at the facility, to sketch or photograph any portion of the facility, to copy or photograph any document or records necessary to determine compliance, or to interview any employees or representatives of the owner, operator or registrant. The New Jersey Department of Environmental Protection (DEP) may enter and inspect premises without any prior warning. DEP need only present appropriate credentials and follow applicable safety procedures. Owners or operators, or any of their employees, must assist with all aspects of any inspection. Assistance includes making available sampling equipment and sampling facilities for DEPE to determine the nature and quantity of any air contaminant emitted. If a business refuses access to an air quality inspector, that inspector may be able to obtain a warrant to enter and inspect that facility.

Inspection schedules vary depending on the emission levels and complexity of the equipment being inspected and on the compliance history of each firm. Inspections are also made in response to complaints. Air districts often conduct unannounced inspections.

An air quality agency may suspend or revoke a permit for any of the following reasons:

- failure or refusal to provide information, analyses, plans or specifications requested

- violations of rules, orders or regulations

- failure to correct any condition when required by the air district

- fraud or deceit in obtaining the permit

Generally, the easiest reinstatements are available when information, analyses, plans or specifications are provided by the source operator. Instances of fraud or violations having to do with equipment changes may require re-permitting.

Air districts generally prefer to encourage voluntary compliance. Therefore, districts use outreach and education rather than fines. Consequently, many air pollution inspectors are encouraged to be as informative and helpful as possible while they are onsite. In the event of a procedural or record-keeping violation, most air districts issue a **Notice to Comply** or a **Notification of Violation**. Every inspector who issues either of these notices is required to follow up to ensure that compliance has indeed been achieved. The first question you should ask an inspector after receiving a Notification of Violation or a Notice to Comply is: "How long do I have to correct this violation?" Violations are usually required to be corrected within a set time--typically from 14 to 30 days. **Texas** inspectors allow 30 days for compliance before issuing an Order for Compliance. Once issued, an Order for Compliance requires compliance within 180 days after receiving a Notification of Violation. If you do not comply by the date given on the Notice to Comply, a district may issue a **Notice of Violation**. This typically means that you are in violation of either district rules or state air pollution laws. If nothing is done to correct the violation, the problem may be handled as a civil case with potentially substantial fines, through a nonjudicial settlement process, or as a criminal misdemeanor.

To the extent possible, businesses and other air pollution sources should **take immediate action to correct the violation**. In some states, many violations can be resolved by mail through administrative provisions similar to SCAQMD's Mutual Settlement Agreement Program.

EPA received several new or strengthened enforcement powers under the 1990 CAA amendments. Under the new amendments, EPA may issue **administrative compliance order** following a Notice of Violation, and can set out a compliance schedule for up to one year (instead of just 30 days, as was previously the case). Knowing violations of a compliance order are now made felonies under the 1990 CAA amendments. In addition, EPA now has authority to issue administrative penalty orders without going to court (where EPA must rely on the U.S. Department of Justice). The practical effect is that only the largest cases go to court; the others proceed through an administrative law judge. EPA can also issue field citations ("clean air traffic tickets") in amounts up to $5,000 per day of violation.

The harshest penalty is 15 years' imprisonment for "knowing" releases of HAPs which may cause imminent danger of death or serious bodily injury. For lesser but "knowing" violations of the act, Congress increased former misdemeanor penalties to felonies, leading to up to five years in prison. Obstructing any of the act's regulatory provisions may be punished by up to two years in jail (as opposed to six months under prior law). Failure to pay fees may result in up to one year in jail. Additionally, Congress doubled all sanctions for repeat offenders. Responsible corporate officials must certify the accuracy of their company's compliance status report. Congress has recognized that employees who violate the act in the normal course of their employment, and at their employer's direction, should not be subject to criminal prosecution. Moreover, since audit, compliance, and monitoring programs are seen as desirable, the CAA amendments do not allow the government to resort to criminal sanctions when such programs are carried out in good faith.

Conduct which **contravenes** the CAA's regulatory scheme includes:

- knowing violation of a SIP requirement continued more than 30 days after the violator receives a Notice of Violation

- violation of an administrative compliance order or administrative penalty requirement

- knowing violations of any aspects of CAA, including:

  -- NSPSs
  -- permit requirements

-- record-keeping, inspection, and monitoring requirements
-- emergency orders
-- NESHAPs
-- acid rain or stratospheric ozone controls

• knowing failure to pay fees

The government need only show that the accused was aware that he or she was committing the particular action, not that he or she knew the action was unlawful.

Air Toxics: Emissions

Until recently, the federal Clean Air Act (CAA) has done relatively little to regulate the wide variety of air toxics. The CAA amendments of 1970 required EPA to establish and maintain a list of air toxics which are named as **hazardous air pollutants (HAPs)**, and to set emission standards for several specific sources of such pollutants. The federal regulatory process over the following 20 years was very cumbersome, however, essentially requiring EPA to amass conclusive evidence that a particular air toxic was indeed "hazardous" before regulation could take hold. As a result, in two decades only eight HAPs were designated, and their emissions subjected to **National Emission Standards for Hazardous Air Pollutants (NESHAPs)**.

Once promulgated, NESHAPs were quite strict. No new source of a designated air toxic compound could be built, and no existing source could be modified, except in compliance with the appropriate standard. But the program's overall scale paled in comparison with the scope of the nation's air toxics problem. A few states, like **New Jersey**, developed extensive air toxics

regulatory programs of their own. Then in 1990 Congress decided to act decisively.

The 1990 CAA amendments, represent an important new national effort in air pollution control. Air toxics requirements are distinct from the new federal programs described; many sources are required to comply with rules adopted for both "conventional" and toxic air contaminants. In most states, a single air quality agency administers both conventional and toxic air contaminant programs.

The 1990 CAA amendments made major changes in the federal approach to regulating toxic air pollutants. In these amendments, Congress listed 189 substances as HAPs, while giving EPA authority to review the HAP list periodically and add or delete substances. EPA must develop standards for **Maximum Achievable Control Technology (MACT)** to control air toxics emissions from thousands of sources across the country.

Tens of thousands of sources of air toxics nationwide--existing as well as new, small as well as large--are now required to apply specific technologies to control their ongoing emissions of hazardous air pollutants. Thousands of facilities also have to obtain and comply with air quality permits under a new national permitting program. These new federal regulatory requirements will become increasingly significant during the 1990s as EPA develops its detailed regulations and guidelines. Existing state air toxics control programs will have to be integrated into these new national regulatory mandates.

Through 1990, EPA had set NESHAPs for only eight air toxics:

- asbestos
- benzene
- beryllium
- coke oven emissions
- inorganic arsenic
- mercury
- radionuclides
- vinyl chloride

In general, each NESHAPs applies to an air toxic emitted in a specific context.

The reader should review Part 61 of 40 Code of Federal Regulations (CFR) to ensure that the facility is in compliance with NESHAPs standards. These requirements cover many pages of federal regulations, so you should refer to specific CFR parts applicable to your operation.

In the 1990 CAA amendments, Congress listed 189 additional HAPs to be controlled by EPA (which can add or subtract from this list). EPA is to develop air toxics emission standards for all of these HAPs over the next several years. EPA has developed a lengthy list of nearly 750 different categories of sources that emit HAPs. This list was published by EPA in the *Federal Register* of June 21, 1991. The list of 750 source categories is divided into the following 20 general industry groups:

- agricultural chemicals: production and use
- chemical production and use activities
- fibers: production and use
- food and agriculture
- fuel combustion
- inorganic chemicals: production and use (two groups)
- metallurgical industry (two groups of nonferrous metals)
- metallurgical industry: ferrous metals
- mineral products processing and use (two groups)
- miscellaneous industries
- petroleum refineries
- petroleum and gasoline: production and marketing
- pharmaceutical production processes
- polymers and resins production
- surface coating and processing
- synthetic organic chemicals: production
- waste treatment and disposal

The category for production of synthetic organic chemicals is the largest single source category.

EPA has divided its 20 general industry groups into four "bins" for which it will develop **Maximum Achievable Control Technology (MACT)** standards. EPA will issue standards for these bins in sequential order at two to three year intervals. Over the next decade, therefore, EPA will specify technological control standards for all of the source categories in each bin. As required by the amendments, these standards will require all sources to use MACT to reduce their emission of HAPs.

- **Bin One:**

  The first bin is to cover 40 industry source categories, for which standards were developed by November 15, 1992. These 40 categories include dry cleaners and a new hazardous organic standard (covering production of 400 chemicals). Coke oven standards covering charging, topside, and door leaks were issued by December 31, 1992.

- **Bin Two:**

  By November 15, 1994, EPA is to develop emission control standards for 25% of all the industry categories.

- **Bin Three:**

  By November 15, 1997, EPA must set standards for 50% of all listed categories.

- **Bin Four:**

  By November 15, 2000, EPA must promulgate air toxics emission standards for all listed categories and subcategories.

Existing sources have three years to comply with the MACT standard once it has been defined for that source category. However, a one-year extension may be granted if the permitting authority (EPA or a state agency with an approved permit program) finds the need for additional time to install necessary controls. New sources may be granted a three-year extension under certain conditions.

EPA's first proposed emission control rule under the 1990 CAA amendments is aimed at reducing emissions of perchloroethylene (also called PCE or "perc") from industrial and commercial dry cleaning establishments. As set forth, EPA's rule affects firms emitting more than 10 tons of perc per year-- some 3,700 of the country's 25,200 dry cleaning facilities. This requires cleaners-- depending on their size, and on whether they are new or existing--to control perc emissions by adopting either MACT or less-stringent **generally available control technology (GACT)**. The EPA rule includes some pollution control measures that require a combination of equipment changes, operating practices, and maintenance procedures. Operators are required to conduct a weekly inspection to prevent solvent emissions from broken or improperly operating equipment. Also periodic record-keeping to track the amount of perc used at that facility is required. EPA estimates that dry cleaners currently emit over 92,000 tons of perc annually.

Sources emitting 10 tons or more of perc are required to install MACT--in this case, carbon adsorbers, refrigerated condensers or equivalent devices (approved by the administering air quality control agency). Sources emitting less than 10 tons are only required to implement GACT, which, in the case of dry cleaners, consists of improved record-keeping and "housekeeping" efforts.

INDEX

Printed and bound by CPI Group (UK) Ltd, Croydon, CR0 4YY

03/10/2024

01040335-0018